Deforestation and Perilous Land Degradation

Forester's Premier Role in Saving India's Destiny

A. B. Chaudhuri IFS (Retd.)
Ex-Director, Forest Survey of India

D. D. Sarkar WBFS (Retd.)
Ex-Deputy Conservator of Forests

CBS

CBS PUBLISHERS & DISTRIBUTORS PVT. LTD.

New Delhi • Bangalore • Pune • Cochin • Chennai (India)

ISBN : 81-239-1280-3

First Edition : 2006
Reprint : 2010

Published by Satish Kumar Jain and produced by V.K. Jain for
CBS Publishers & Distributors Pvt. Ltd.,
CBS Plaza, 4819/XI Prahlad Street, 24 Ansari Road, Daryaganj,
New Delhi - 110002, India.   • Website: www.cbspd.com
e-mail: delhi@cbspd.com, cbspubs@vsnl.com, cbspubs@airtelmail.in
Ph.: 23289259, 23266861, 23266867   • Fax: 011-23243014

Branches:
• *Bangalore:* Seema House, 2975, 17th Cross, K.R. Road,
  Bansankari 2nd Stage, Bangalore - 560070   Ph.: 26771678/79
  Fax: 080-26771680   • e-mail: bangalore@cbspd.com
• *Pune:* Bhuruk Prestige, Sr. No. 52/12/2+1+3/2,
  Narhe, Haveli (Near Katraj-Dehu Road by Pass), Pune-411051
  Ph.: +91-20-32404169   • Fax: 020-24464059
  e-mail: pune@cbspd.com
• *Cochin:* 36/14, Kalluvilakam, Lissie Hospital Road,
  Cochin - 682018, Kerala • e-mail: cochin@cbspd.com
  Ph.: 0484-4059061-65   • Fax: 0484-4059065
• *Chennai:* 20, West Park Road, Shenoy Nagar, Chennai - 600030
  e-mail: chennai@cbspd.com   Ph.: 044-26260666-26202620
  Fax: 044-45530020

Printed at :
India Binding House, Noida (UP)

to

D. Brandis, W. Schleich, B. Ribbentrop

and

generations of foresters for

their yeomen services for

conservation of

India's

forests and environment

Acknowledgements

The authors express their heartfelt gratitude to the PCCF and ACCF, WB and others concerned for allowing them the use of the Forest Library at Aranya Bhaban and to Smt. Sima Ghosh, Librarian-in-Charge. They offer their sincere best wishes for helping them ungrudgingly in the use of the library.

The authors also acknowledge with utmost gratitude for the untiring effort of Sri Dipyaman Sarkar for doing the secretarial jobs, drawing of sketches and computerising the entire manuscript including the DTP work and final composition.

Foreword

Mʀ A.B. Cʜᴀᴜᴅʜᴜʀɪ has chosen a vital present day issue in his research treatise. This shows his ingenuity and how deeply he is concerned about the very existence of this country. He firmly believes India stands on an unstable foundation that is sure to crumble bringing a colossal disaster to entire country. In recent years various media reports, besides the vociferous claims of the environmentalists and NGOs about the perilous landuse situation and anthropogenic and other biotic problems which are sure to nullify all development efforts. Chaudhuri feels our government is playing similar role as the emperor 'Nero' of Rome who was playing his fiddle when Rome was burning.

My knowledge of forestry is not extensive, but my long years of experience in civil administration and judicial arena helped me to be acquainted with human and environmental interactions and issues that lasts over sixty years. I was a student of chemistry and had the privilege of working under the famous scientist late Dr. K. Khuda. This helped me develop an analytical bend of mind that helped me to judge man environmental issues all through my service tenure. I have been a next door neighbour of Shri Chaudhuri and had the opportunity to read a number of books written by him on Plant, Wildlife and Environment which helped me widen the sphere of my knowledge. His lifetime contribution to botany and plant science has been widely acknowledged.

Three premier issues precipitate in his book which are: (i) India's 'doomsday' is knocking at the door which may be read from the facts and figures presented by him about soil erosion, flood, drought, siltation, forest denudation, etc. as detailed in Chapter 6, (ii) annihilation of forests and wildlife resources which goes unabated on which Chaudhuri presented several recent media reports as may be read in Chapters 5, 8 and 9, and (iii) his sincere solution of this disastrous situation is in observance of a 15 year's 'Moratorium' as detailed in Chapter 16.

In his book Chaudhuri begins with an invocation to "Nature-God" and says how the forester's lifelong activities are interwoven with the rhythm and balance of nature. He also deals with present the day issues like eco-development and sustainability.

Chaudhuri has drawn a broad outline on the history of Indian forest administration and conservation. He mentions briefly the foresters' training and various other creative activities. It is a noble profession aimed at selfless services for conservation of resources of the country. He suggests a new curriculum for country's 500 million school–college youths which should be oriented towards love for plants, and practical lessons in afforestation. The President of India Dr. A.P.J. Abdul Kalam's Independence Day speech of 2005 aims to massive afforestation to mitigate unemployment problem, and building up a strong resource. Chaudhuri seeks a solution in the line of USA President (Late) F. Roosevelt (1931) of USA who created an emergency service called CCC (Civilian Conservation Corps) and solved the problems of erosion, flood, drought, unemployment in the USA.

Chaudhuri has designed the book for school-college students and general public; not for expert foresters or high profile readers. I wish his dream will take a shape and a 'New India' will emerge. The country will have a sylvan array through afforestation and soil conservation works and India's 500 million young generations will take a vow to build a strong character through man making education and shall not indulge in cheap extravaganza and fun frolic. Otherwise the mighty Himalayas will crumble on the vast plains and E1 Nino or Tsunami like disaster will engulf the entire country of the *Ramayana* and the *Mahabharata*.

B.M. Mazumder, M.Sc., LLB.
Addl. Standing Govt. of India Counsel
131, NSC Bose Road, Block 9, Flat 8
Kolkata 40

Preface

It was the midnight of December 25, 2004, the senior author was enumerating the extent and size of biotic and abiotic calamities brought about by the various agencies like flood, drought, landslides, erosion, soilwash, siltation, forest depletion, extension of drought-prone areas, etc. As the night progressed, the TV flashed the news of 'Tsunami' tragedy. Amidst distress and pain the author breathed a sigh of relief as his apprehension was a realistic one and 'Tsunami' was just an additional evil he had listed. A few months back some experts pointed out 'El-Nino' like environmental menace causing drought in Orissa. Earthquakes have proved disastrous in Gujarat, Maharashtra and North-East India.

But the planning of India were done without considering the horrendous landuse situation on which the country's future stands. It is like building of castles over a soft, cracked, dilapidated foundation that might break away like quick sand. The authors drew a parallel between the Roman emperor Nero (37–60 AD) (dreaming of building a new city of Rome while he was playing his lyre in full view of burning of the city), and the planning for development of modern India; the latter having been fully oblivious of landuse situation.

Besides these natural calamities India has to tackle the problems of a huge poverty-stricken, illiterate, unemployed mass where even amenities like drinking water, medical help, etc. are lacking.

The authors, while analysing various facts and figures, documented by the Planning Commission of India, the National Commission on Agriculture and the National Commission on Flood found that they have registered about 70% of the total 3.2 m sq. km of land has been reeling under flood, drought, erosion, etc. Such terrible landuse condition has to be tackled effectively before India takes up massive, gigantic and long duration projects. They have stressed that India needs to create an effective cover of plants by massive afforestation mingled with soil conservation engineering works, digging up of ponds and cleaning them of siltation. Here comes the unique role of the foresters who are well-equipped to create a better India. They have training, expertise and the character, built over an experience of about 200 years to rehabilitate depleted soil.

The book has been designed to serve and convince the intelligentsia that the foresters create and protect human destiny and their prosperity; when a forest fails to regenerate naturally and gradually fades to extinction over the years, the foresters come to the rescue. They artificially procreate by raising nursery and taking meticulous care and plant them to create a forest. So the foresters serve posterity by providing them not only their requirement of fuelwood, timber, fodder, fruit, etc. they thereby create a salubrious climate and a pollution-free landscape.

The forests during the process of development immediately attract ground and arboreal animals and birds, reptiles, lizards and other myriads of species of animals gradually taking shelter and for food and living. So, by planting trees the foresters :

- Gift a valuable wealth to posterity;
- provide shelter to innumerable species of macro and micro animals;
- create a pollution-free healthy climate;
- provide recreational sites for the tourists;
- create a natural laboratory and museum for students of science to work and learn;
- and many other benefits accrued from the forests.

Kolkata has been marked as the world's third most polluted city. Winter air is the foulest in India, 10–12 lakh people have respiratory diseases. In the past decade, 50% rise in respiratory diseases (*The Telegraph:* Dec, 2004) have been registered. For centuries the foresters are in the service of humanity. But India needs more areas to be covered with vegetation to fight similar hazards. India needs dense tree planting in mine areas and industrial sites in particular besides various other vulnerable areas.

A fresh terror in global warming, has been indicated by a UN organization for security and cooperation in Europe (OSCE) which points out at the ecological roots of conflict in the tension ridden Africa in land degradation; also rising of sea level may force millions of Bangladeshis migrate into India, fuelling ethnic and religious tension that end in bloody riots (Reuter, *The Telegraph* daily, dt. 24.10.2004). India's land degradation is an additional calamity.

This treatise opens with an "Invocation" to "Nature God" and the "Rhythm" in forests. Nature and rhythm in forests are interwoven with the life and activities of the foresters. They are part of the nature. Various qualities and functions of forests have been brought to light by charts. The qualities and activities of the foresters also have been presented in chart-form. The landuse situation has been drawn elaborately in Chapter 6 to show that India faces a catastrophic future and that the forests and foresters can save the country from destruction. So, to begin with, massive afforestation and soil conservation works have to be given top priority and later planning for other development works may be taken into consideration progressively with prevailing conditions.

The present treatise of 300 pages can hardly relate the elaborate history of forest conservation and administration as it will need about thousand pages even to draw an outline of the history of forests and the foresters. The readers must not expect various issues that the foresters are concerned with today and the activities performed by them in this short presentation.

Forestry is one of the noblest professions on earth and foresters, "The Angels", are meant to perform to create a better environment, for playing constructive role and protect humanity from destruction of forest.

Three brilliant officers who shaped Indian Forestry in various fields are Sir (Dr.) D. Brandis, Dr. W. Schleich and Dr. B. Ribbentrop. They had vision and a profound knowledge on a wide variety of subjects. It is because of the foolproof recruitment systems introduced by Dr. Brandis, the father of Indian Forestry, that the forest service could create a tradition of a large group of sincere efficient officers of strong character. Dr. Brandis felt that a forester should have tact, a sense of sound judgement, high moral character, good family linkage, even temper, good health, super abilities, humane in dealing with local people.

For this rigid recruitment procedures, the forest service personnel all over India had a distinct tradition and identity other than even disciplined army and police personnel. This tradition has been maintained over the last 200 years.

However, the people in general and the intelligentsia in particular know little about forest personnel and the nature of their duties and that they have to perform multifaceted duties. The foresters administer 20% of the country's land with bare minimum resources and amenities.

All planning efforts to build up a better India will not yield any fruitful result due to the poor and unstable foundation on which our country stands. Long ago Swami Vivekananda visualised that the youth of the country should possess strong will and determination, based on 'man-making and character-building' system of education; besides the villagers must be capable of improving their own fate by hard work, not depending upon outside help. What we find today, even after more than half a century of independence, that the dream of the sage of sages – *Swami Vivekananda* remains unfulfilled.

India needs adoption of 15 years' 'moratorium' to build a strong foundation on which the country's superstructure of modern India can be built. For this India needs:

- Cutting off all extravagant expenditures of colourful ceremonies, film shows, beauty contests, car racing, expensive rituals, luxurious travels, building luxurious hotels, savoury cuisine.
- School/college curricula must contain syllabus highlighting the catastrophic landuse situation of the country and the need for massive afforestation and soil conservation works. The students should know about the socioeconomic backwardness, political instability, poverty, illiteracy, primitive social customs, drought,

flood, etc. of the country and the usefulness of forest environment. The entire education system should be based on the 'man-making and character building' – system of education.

- An effort similar to Wangar Mathai of Kenya to protect forests (this fetched for her Peace Nobel Prize of 2004) is necessary;
- India has to build up a department on the line of Civilian Conservation Corps (CCC) which the late president F. Roosevelt of USA conceived in 1931 to tackle the problems of unemployment, erosion, drought, wind storm, soil wash by massive afforestation and soil conservation programme.

This suggestion gets support in President of India Dr. A.P.J. Abdul Kalam's Independence Day (2005) deliberation for an action plan for rural employment generation to help achieve 10 percent GDP growth rate. Dr. Kalam said (based on Planning Commission's figures) that the objectives will be achieved through employment generation of 76 million people in rural areas for which he suggests Bio-fuel plantation, waste land development, water harvesting, bamboo plantation, and the like.

The people should temporarily forget the comfort of luxurious extravaganza of five and seven star hotels, multiplex, sophisticated and expensive private and official functions, abandon cheap pop-song, burning of petrol/diesel in activities other than essential services, more liquor shops, biting of Mughlai, Thai, Afghan, Chinese, Tibetan and Continental cuisine and sucking guzzler's delight, Bratwurst (Chicken Pork Sausages), Brathend (Chicken Roast), Leberkaese (meat loaf with fried onions and egg) and the like.

Even at this critical sociopolitical and socioeconomic unstable situation of the country a good section of intelligentsia (connoisseur of culture of the country) dream of whole night rendezvous round the year, the sights and sounds of city's festive fever, the flavours of thousand savoury and sweet delight, roam about from pandal to pandal in quest of cultural evolution, prefer fun and frolic, being oblivious of critical need of the country for which Mother India is in need of a great sacrifice. The entire length of country's border is teeming with criminals, local gangsters and is threatened with intrusion.

At this juncture afforestation and soil conservation issues should be considered as national issues and must be tackled on a war footing. Our school and college students are environment conscious on paper only and they should actively participate in such programmes.

Some experts feel that the trend of environmental degradation indicates that several departments should be brought under one umbrella and a firm policy and action plan developed, followed by a strong political will to save the country from the present disaster.

Planning Commission of India (1982) indicated that India can achieve ecological security by increasing vegetal cover to tackle the problem of serious degradation in various fields.

The authors sincerely feel that Indian Foresters are sure to take a leading role in shaping the country as proposed as they have 200 years' tradition of conservation and afforestation work.

B. B. Vohra, ex-chairman of Flood Commission of India, in his report suggested that there should be a Ministry of Land Management and creation of a National Land Development Board to finance various land improvement projects. Mobilization of students should be most useful for large scale soil conservation and afforestation programmes. There is immense scope for voluntary organizations to involve themselves in these activities. Vohra identifies renovation of India's 5,00,000 odd tanks and 900 odd project reservoirs and denuded watershed of India's flooded rivers. Mention may be made of ITC's one of the Social Forestry Programmes in Chetavarigudam village of Khammam district of Andhra Pradesh, where it used high-yielding disease resistant, clonal saplings developed by it at its state-of-the-art biotechnology research centre. More than 26,500 hectares of wastelands had been rejuvenated with 108 million saplings, creating livelihood opportunities for nearly 2,00,000 people. This gives a positive guide to the future livelihood and employment of a vast population to be engaged in afforestation and soil conservation works all over the country.

The need for saving India's dwindling forest cover has prompted Indian Foresters to undergo management courses. Forest Management is all about conservation and maintenance of the forest wealth and to administer the forests. The work implies application of technical forestry principles, practices and business techniques, hence, forest management involves the practical application of scientific, economic and social principles of forests. Foresters manage a forest, keeping some multifaceted objectives in view some of which are too maximise production of timber, fuelwood, furniture wood, paper and pulpwood, fodder, and to meet the local demand of the people, exploitation, creation of plantation and also management of wildlife. Every state has its own management organisation and the forest policy provides conceptual guidelines on forestry operations. The administration includes analytical and managerial skills, application of concepts to managerial problems, synthesis, technical skills, fieldwork, basic accounting, management accounting, financial management and marketing and project management. Forest management in fact means the accomplishment of multiple goals such as sustainable forest management, livelihood issues in rural India, community forestry, gender issues, legal policy analysis in forests, protected area and biodiversity conservation, management of ecosystem and watersheds, soil and water conservation, maintaining biodiversity and hydrological cycles and carbon sequestering. So the management issue involves judicious combination of management, service and forestry sciences besides several others.

xiv Deforestation and Land Degradation

In the event of perilous landuse situation worsened by forest depletion, fire damage, grazing menace, encroachment and illicit felling which started a thousand years back and are still continuing, have made the management a tough and challenging task.

During the last quarter of twentieth century new concepts such as social forestry, farm forestry, joint forest management, eco-development and eco-tourism, etc. have caught up in a major way. Forest Development Corporations have been setup in all states for quick conservation of forests, marketing of timber and other forest produce, and large scale afforestation. Geographic Information System (GIS) is being used now to map and monitor forest cover. Satellite imagery interpretation is also being widely used. However, detailed discussion of all these issues do not come under the purview of this treatise.

All these issues find a place in various chapters though presented very briefly. The observations made in this treatise, it is believed, will motivate the builder's of the nation to guide the unemployed youth, students and others to be involved in massive work of afforestation, soil conservation, shelter belts, parks and gardens, coastal and canal bank plantations and scores of such activities. In a span of 15 years the country would surely to turn green.

The authors categorically want to mention that the book is designed for school and college students, besides the vast mass of people interested in forests and their impact on the environment as a whole. The future generation readers will surely appreciate the nature of training imparted to the foresters, which moulds an individual to a real man. They also feel that the entire education system should be modified in the similar line to create new generations of 'mens sana incorpore sano' (a sound mind in a sound body). They feel the urgency of similar rigorous training to be imparted to the teachers in general and primary teachers in particular and also to the officers of all civil departments.

A short chart is given on page *ii* for representing the contents of the book at a glance.

A. B. Chaudhuri
D. D. Sarkar

131, N. S. C. Bose Road,
Block 10, Flat - 4
Kolkata – 700 040

Contents

been assigned with some works in D.T.R.

Chapter 4
A Well-protected Forest Creates Environment

Chapter 5
Conserving Biodiversity Resources

Chapter 6
India is Facing a Perilous Landuse and Environmental Crisis

Chapter 7
Forest Tribal Areas and Pollution

*Tribals are Exposed to Forestry and Mining Operations and
Pollution Therefrom* 131

Chapter 8
Foresters are Pioneer Ecologists

*Foresters were Pioneers in the Application of Ecological
Principles and Theories* 149

Chapter 9
Management of Wildlife Sanctuaries in Aquatic Sites

Forester's Achievements 159

1

Forests are an Integral Part of Nature

— *Foresters Adore Nature God in their Activities*

*T*his treatise opens with an 'Invocation' to 'Nature God'. Nature's various processes are interwoven with the life and activities of the foresters. They are the part of the nature. Various qualities functions of forests have been brought to light in the form of charts. The qualities and activities of the foresters also have been presented in chart-form.

Various quotations presented from Barbe Baker's book, 'Green Glory' give vivid pictures of beneficial role of forests and the evil effects of forest depletion.

Going through the papers the readers may visualise the great heights, great depths, light and shade, peace and conflict, facts and fancies—all curiously entwined in perfect balance. Also of myriads of living communities throbbing with life and abode of mysterious forces.

"Here war and peace, life and death, silence that is punctuated by a thousand sounds, the faint rustling of a leaf, the shrill shriek of a bird, or the furtive movement of badger. A call or the cry of a bird comes back with hallow echo as through the trees took up the chorus".

1

Forests are an Integral Part of Nature

— Foresters Adore Nature God in their Activities

FOREST AND FORESTERS

Foresters are wedded to nature. They stay inside the forests, spend their entire service tenure engaged in various activities inside the forests; they breathe, sing, dance, walk and think in the rhythm of forests. They carry with them a pure and serene nostalgic memory of forests when they retire from services for they have faced various hazards in performance of their duties as they hoped the posterity would harvest all the benefits from the forests they have created and protected.

The author of this book are a part of the group of dedicated "Creator of the destiny" for the future generation and pray to God that all their masterpieces of creation on flora and fauna – lasts for eternity.

Let the foresters' creation become a paradise for adoration of 'Nature – God' for generations to come. People know that forests are community of living trees and associated organisms; utilizing sunshine, air, water and earthly materials they attain maturity, reproduce itself and are capable for furnishing mankind with indispensable products and services.

PLANTS – ARE INTEGRAL PART OF NATURE

"The lovely form of a tree or a leaf, the galaxy of colours, the brightness of the black thorn, a bed of primroses, or a carpet of blue bells, the mystic, silence broken only by the song of a bird or the gentle stirring of the breeze as it caresses the tree tops – here are things which rest, quicken and then stimulate the mind and spirit of man. Here is a quality and power which neither biology, physiology nor any other 'Ology' can explain. Here is beauty; and no man can gaze on beauty without appropriating something of it to himself – unless his sense of what is good and lovely has been smothered beyond hope of redemption".

[*Source:* **Our Forest by W. H. Rowe**]

NATURE : MY GOD (A forester's invocation)

RHYTHM OF LIFE

I have discovered the rhythm, balance and life processes that exist in nature. I have felt the intricate and complex interactions that flow on between plant, animal and man. The forests of India offer such inexhaustible facets of nature.

In the language of a few authors I also express my sincere oneness of feelings as below when I –
Sing the rhythm of life,
Dance in the rhythm of life,
Breathe in the rhythm of life,
Think in the rhythm of life,
Walk in the rhythm of life.
In their tunes I also sing –
The beauty of flowers that draws my heart and exalts my soul,
Stillness of the night fill and engulfs me with wonder.
I sing about the eternity of nature and its immortal creation as the –
"Beauty of flowers draws my heart,
Beauty of dawn exalts my soul,
Stillness of the night fill me with wonder,
Beauty of face and beauty of sight
Awaken in my soul love and ever
                    greater love".

[Source: **Unknown]**

My heart is awed within me, when I think
Of the great miracle that still goes on –
In silence, round me – the perpetual work
Of thy creation, finished, yet renewed
For ever.

[Source: **Bryant]**

BALANCE OF NATURE

A forester helps nature and work with her; and Nature will regard thee as one of her creators and make obeisance. And she will open wide before thee. The portraits of her secret chambers lay before thy gaze the treasures hidden in the very depth of her virgin bosom.

[The Voice of Nature]

FORESTS ARE ETERNAL

An age builds up cities,
   an hour destroys them.
In a moment the ashes are made,
   but a forest is a long time growing

[*Source:* **Seneca – from Latin**]

FORESTS AND LIFE'S RHYTHM

"Walk in the rhythm of life;
Your limbs will not tire.
Sing the rhythm of life;
Your voice will gain sweetness.
Dance in the rhythm of life;
Your feet will not touch the ground".

[*Source:* **Modified from Swami Paramananda of
R. K. Mission, Chennai**]

FORESTERS – FEEL ONENESS WITH NATURE AS THEY

"Breathe in rhythm.
Think in rhythm.
Talk in rhythm.
Walk in rhythm.
Sing in rhythm.
Dance in rhythm" which is a part of their lives.

[*Source:* **Modified from Swami Paramananda of
R. K. Mission, Chennai**]

GLORY OF TREES (Trees are eternal gift of God)

Foresters sing the glory of God in the tune of H.V.D. Ke as below:
"But the glory of trees is more than their gifts;
'T is a beautiful wonder of life that lifts,
From a wrinkled seed in an earth bound clod,
A column, an arch in the temple of God,
A pillar of power, a dome of delight,
A shrine of song, and a joy of sight;
Their roots are nurses of rivers in birth,
Their leaves are alive with the breathe of the earth;
They shelter the dwellings of man; and they bend,
O'er his grave with look of a living friend."

[*Source:* **Modified from Henry Van Dy Ke**]

TREES ARE CONSTANT COMPANION OF FORESTERS

a praise
"I think that I shall never see
A poem love as a tree
A tree whose hungry mouth is prest
Against the earth's sweet, flowing
A tree that looks at God all day
And lifts her leafy arms to pray;

A tree that may in summer wear
A nest of robins in her hair;
Upon whose bosom snow has lain,
Who intimately lives with rain,
Poems are made by fools like,
But only God can make a tree."

[*Source:* Joyce Kilkmer – 'Trees']

RHYTHM AND RHYTHMICITY IN FORESTS

"Functional rhythm is the basic property of living thing, i.e. of protoplasm. All the vital activities of living organism follows a rhythmic sequence. The metabolic activities connected with the very existence of the organism, like nutrition, respiration, etc. have a period of tense activity (active phase), followed by a requisite lucid gap (passive phase). These active and passive phases rotate cyclically that attributes a distinct rhythm to the activities and stands as an outstanding character of a living matter."

FORESTERS FIND THE

- rhythm in ebb and nip tide that goes on an on in the estuaries
- rhythm that works in enstuarine animals as "biological clock"
- rhythm in the rising and setting of sun – at a forest background
- rhythm in the migration of honey bee and birds in forest every year at a particular time
- rhythm in the migration of bird in their breeding season to a particular site
- rhythm in the long distance migration of turtles in the coastal areas for breeding
- rhythm in the unfolding of leaf and floral buds
- rhythm in the loud monotonous ringing 'tuk-tuk' repeated call of barbets
- rhythm in the movement of earth round in axis to give phases of light and darkness
- rhythm in the movement of earth round the sun above the orbit that brings in the changes in seasons.

THE NATURE GOD (in Forest)

Poets, Philosophers and Intellectuals conceive her as we also do, a few such instances are:

Albert Einsteine: "Look deep into nature and then you will understand everything better."

Michael da Montaigue : "Let us permit nature to have her, she understands her business than we do."

H. D. Thorean : "It appears to be a law that you cannot have a deep sympathy with both man and nature."

Chief Seattle : "Humankind has not woven to the web of life. We are but one thread within it; whatever we do to the web, we do to ourselves. All things are bound together."

Goerge Washington Carver : "I like to think of nature as an unlimited broadcasting station, through which God speaks to us every hour, if we will only tune in."

William Shakespeare : "One tough of nature makes the whole world kin."

Luther Burbank : "If you violate (Nature's) laws – you are your own prosecuting attorney, judge, jury and hangman."

Carl Sagun : "The Universe is not required to be in perfect harmony with human ambition."

John Dryden : "God never made his work for man to mend."

Rene Dubos : "Nature always strikes back. It takes all the running we can do to remain in the same place."

Cummings : "The world is mud-luscious and puddle – wonderful."

Aristotle : "If one may be better than another, that you may be sure is Nature's way."

John Muir : "When one tugs at a single thing in nature, he finds it attached to the rest of the World."

John Keats : "The poetry of the earth is never dead."

Lord Byron
"There is pleasure in the pathless woods,
There is a rapture on the lonely shore,
There is society. when none intrudes
By the deep sea and music, in its roar
I love not man the less, but Nature more"

Alfred Tennyson
"Any man that walks the mead
In bud, or blade, or bloom, may find
A meaning suited to his mind."

Franklin Roosevelt (President of America) : "Back to the land with forestry – without vision a people perish."

"The fundamental idea of forests is the perpetuation of forests by use. Forest protection is not an end of itself; it.is a means to increase and sustain the resource of our country and the industries which depend upon them."

MAGICAL QUALITIES OF FORESTS (As the foresters are aware of)

"Forest (and nature) are composed of strange contrasts : great heights and great depth, light and shade, peace and conflict. Fact and fancy are curiously entwined so that the one is incomplete without the other."

"The forest is a living community throbbing with life – seen and unseen – the abode of mysterious forces. Here are war and peace, life and death, silence that is perpetually punctuated by a thousand sounds, the faint rustling of a leaf, the shrill shriek of a bird or the furtive movement of a tiger. A call or the cry of a bird comes back with hollow echo as though the tree look up the chorus; the forest lives."

"Here the strange flickering light of the sun gleaming through a space in the leafy canopy casts a shadow which suddenly springs to life in the form of a rabbit battling for friendly shelter. A squirrel makes a home – may a world – of beech tree while a bird wages war upon lesser forms of life within this community. For here, indeed is a colony of individuals – plants, trees and animals – coexisting in a communal life wherein strife and mutual aid go constantly together."

[*Source*: **Modified from "Our Forests"
by W. H. Rowe**]

FOREST COVER IN TOTALITY MAY SAVE INDIA FROM DISASTER

Fragile foot hill swalicks in the geographically recently formed Himalayas, impact of heavy population, poverty of the people, poor supply of non-wood energy materials, poor forest cover over vast areas, annual and ever increasing devastating flood and drought, lack of drinking water and irrigation water facilities and several other such factors have posed serious problems and there is no sign for early recovery from these disasters and survival however we may bloat in vain of a resuscitated India in a few fields. People are aware that the forests have a very material effect upon climate, water supply and soil fertility. Temperature cools down within a wood. The cool air rising from a large woodland area tends to condense its moisture in the form of mist and rain while warm air from bare and barren land disperse.

Trees are continually taking in water from the soil and transpire it again into the atmosphere. A woodland has a great moisture absorbing capacity. Heavy rain falling upon a bare hill side tends to flow quickly towards the valley, carrying with it some of the surface soil. Water absorbed in a woodland goes deeper into the soil gradually.

The disastrous annual flood that runs on many parts of India would not have occurred if the country was well wooded and had a thick ground cover of plants. The washing away of fertile soil by erosion has been a serious menace. The seriously depleted plant cover of catchment of rivers is another serious matter. Sudden flooding and landslides are regular phenomena.

MAN AND THE TREE

"Each of us is a tree. We stretch our arms if we are trees – while our feet are solidly grounded in the earth as the root of the trees are ...."
"Forests are the temples
Trees are the alters
We are the priests serving the Forest Gods
We are also the priests serving inner temple
Treat yourself as if you were an inner temple
To the God which resides within
To walk through the lip as if you were
In one enormous temple
This is the secret of grace."

[*Source:* **Richard Barbe Baker in
My Life, My Trees, 1979**]

"Forests and spirituality are intimately connectad." (anonymous)
"Forests' are Sanctuaries. To see a world in a grain of sand and a heaven in a wild flower."

[*Source:* **William Blake**]

"Intricate are the relation between a single true and the forms of life that live in it and around it."

(Anonymous)

"I shall ever try to drive all evils away from my heart and keep my love in flower knowing that thou hast the seat in the innermost shrine of my heart."

(Rabindra Nath Tagore)

2

Green Glory
— A Brief Presentation on Forests and Environment

In the language of Richard St. Barbe Baker, this chapter presents briefly the protective and environmental roles of forests. He explains in simple language all the catastrophes faced by India citing examples of various countries of the world. Such examples cover ruthless uprooting, burning and pilferage of forest resources, and erosion of soil, resulting in floods and drought in various countries. The author also sings the song of rhythm in forests and its various protective and productive roles.

2

Green Glory
— A Brief Presentation on Forests and Environment

Richard St. Barbe Baker's book *Green Glory* the story of the forest of the world (1948) is perhaps the best treatise in the protective and environmental roles of forests. B. Baker has in simple language expressed all the catastrophes faced by India which have been enumerated in the present book. The authors find it very relevant to present most of the chapters from Barbe Baker's book with some of Baker's own observations as follows.

R. Barbe Baker teaches us that is "All flesh is grass." "Thus the tree, with the help of all plant life, controls the food supply and life of man and of the animal kingdom".

He gives terrifying pictures of man flying in the face of Nature, ruthlessly uprooting and burning the very stuff that holds the world together, and no less terrible pictures of nature making her implacable reply. Man striped the forests of China; Nature swirled away in the yellow river every year 2,50,000 tonnes of the soil on which man might live. Man strips the western prairies to the bone; Nature hands him a dust bowl.

About the Gold coast, Baker observes about the destruction of forests – "Here may be witnessed racial suicide on a bigger scale than the world has ever before seen. Knowing the end of the forest to be near, and with little chance of getting food, the chiefs have forbidden marriage, and the women refuse to bear children, for they will not save sons and daughters from starvation."

The authors feel that Barbe Baker's book *Green Glory* should be included in the school and colleges curricula all over India.

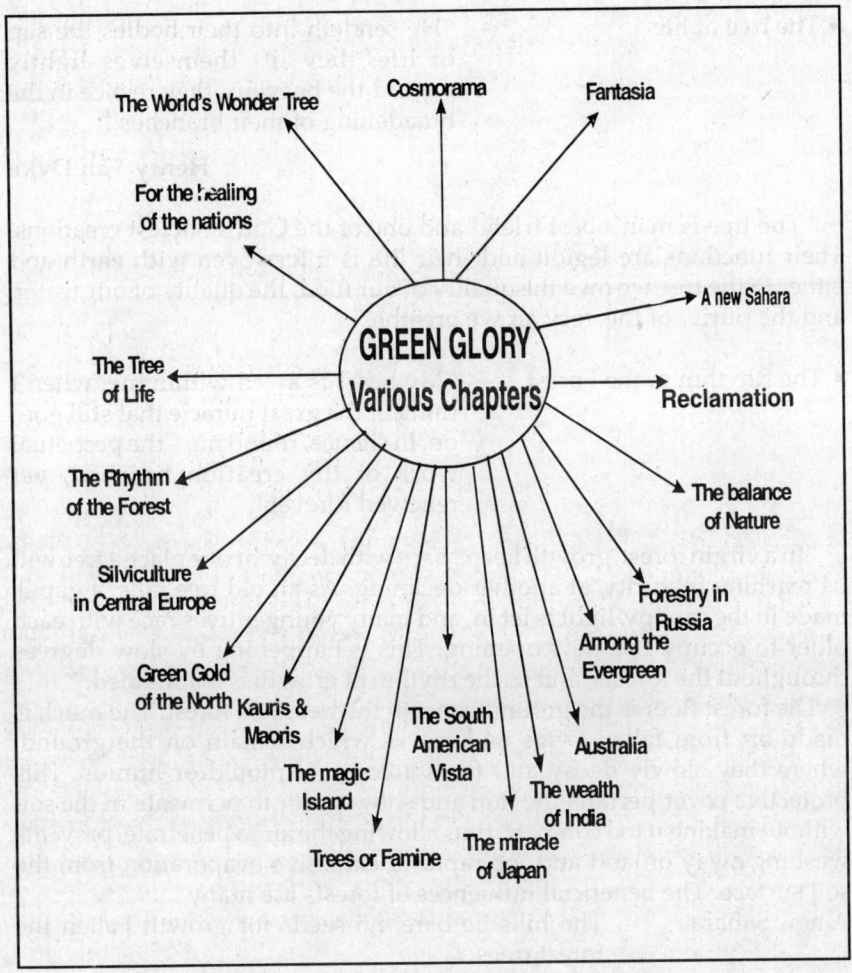

Barbe Baker's Observations

- Cosmorama
- Fantasia
- The world's wonder trees

- "In the beginning God created the heaven and the earth ... and that it was good." Gen : 1
- "This is the forest primeval." Long fellow
- "I will sing of the bounty of the big trees. They are the green tents of the almighty, he hath set them up for comfort and for shelter."

Henry Van Dyke

- The tree of life   – "He sendeth into their bodies the sap of life, they lift themselves lightly toward the heavens, they rejoice in the broadening of their branches."

Henry Van Dyke

"The tree is man's best friend and one of the God's noblest creations. Their functions are legion and their life is interwoven with earth and either to the tree we owe the quality of our food, the quality of our water, and the purity of the very air we breathe."

- The Rhythm of the Forest   – "My heart is awed within me, when I think of the great miracle that still goes on. In silence, round me – the perpetual work of thy creation finished, yet renewed forever."

"In a virgin forest, growth keeps pace with decay: at one place a tree will be reaching maturity, at another declining. As an old tree falls, a gap is made in the canopy, light is let in, and many younger trees race with each other to occupy this new opening. This is happening by slow degrees throughout the forests, and so the rhythm of growth is maintained."

The forest floor is the ground beneath the trees in a forest. The much is made up from fallen twigs and leaves which remain on the ground, where they slowly decay and form a cover of mould or humus. This protective cover permits the rain and snow water to permeate in the soil without making it too compact, thus allowing the air to penetrate, prevents washing away of land and too rapid or excessive evaporation from the soil surface. The beneficial influences of forests are many.

A new Sahara   – The hills lie bare, no seeds for growth Fallen the forest trees.
And winds, haunting the lonely space.
Mourn for the sons of these.

Edith Livingstone Smith

"Although it has taken close on 2000 years to create the great Sahara of Africa, it has taken but two centuries to form the Dust Bowl of America."

Reclamation

"The fundamental idea of forestry is the perpetuation of forests by use. Forest protection is not an end to itself, it is a means to increase and sustain the resources of our country and the industries which depend upon them."

Franklin Roosevelt

When Franklin D. Roosevelt became the President of the USA in March 1933, he assumed office with the country apparently heading for disaster.

Roosevelt on Civilian Conservation Corps

On March 22, 1933 Robert Fechner received a long distance call from the White House requesting his immediate presence in Washington. That very evening a Bill had been sent to Congress by the President, proposing the setting up of a vast army as a measure, partially to relieve unemployment and, at the same time, to accomplish useful work. Congress enacted the legislation, and the President signed the Bill on March 31. On April 5, Robert Fechner was appointed Director of Civilian Conservation Corps.

The creation of the Civilian Conservation Corps was stupendous task, but the genius of the President inspired the cooperation of the four departments – War, Interior, Agriculture and Labour – whose united efforts achieved unprecedented success.

Innumerable camps were established to house 2,50,000 young men, experienced war veterans and work of planting trees began. Next year 3,00,000 men started working in C.C.C. They carried out reclamation and rehabilitation work. They drained ditches, built check dam, planted quick growing species and vegetation to protect farmland from soil wastage to conserve water and prevent flood. They campaigned against tree diseases, improve conditions for wildlife and host of other works related to the national task of conserving and rebuilding America's natural resources. C.C.C did commendable work on fire protection, land reclamation, prevention of soil erosion, wind break, etc.

The launching of the C.C.C. inaugurated a nationwide war of reconstruction – not a war of destruction. It was waged to build up America's human and natural resources. On the other hand President Roosevelt through the C.C.C. camps, was able to save an army of idle youth from the moral erosion caused by unemployment, while on the other he attempted to save his land from deterioration and barrenness.

By launching the C.C.C. he created a new frontier for them, and hundreds of thousands of these young men, who otherwise have had little or nothing to do, were sent to outdoor camps in the forests and parks, where they were given jobs recreative, jobs in the fullest sense of the word. They were taught self discipline, not military training. This army in turn cooperated with Nature in helping to restore the green glory to the hills and fertility to the valleys and plains.

The Balance of Nature

"Help nature and work with her, and nature will regard thee as one of her creators and make obeisance. And she will open wide before thee petrels of her secret chambers, lay bare before they gaze the treasures hidden in the very depths of her virgin bosom."

The Voice of the Silence

- Earth's green covering is the basis of life.
- Trees are the highest example of the vegetable kingdom. In the forest the process of growth always balance one another.
- Silviculture in Central Europe – Silviculture is the art and science perpetrating the woods and forests for the service of man for all time. German is known among the foresters the fatherland of forestry.

Green Gold of the North

We had better be without gold than without trees.

John Evelyn

Sweden is the most wooded country in the world. Her (a very small country) export of Forest products, timber, woodpulp and paper is next to Canada (the largest in the world).

Little coal is to be found in Sweden, but she is called the kingdom of **"White Coal"** – the **"White Coal"** which has electrified her rail roads and brought light and power to her farmers, comes as a by-product from the forests. This **"White Coal"** is the power of water. These same waters which bring wealth to the soil pulsate with power, which is harnessed for use of man. The forest is the mother of the rivers, and rivers are the roots of exit and yield the power for industry.

Baker says that much of the work of afforestation in Norway was being carried out by the Norwegian Army, and that during manoeuvres the soldiers were mainly responsible for fire prevention.

Among the Evergreens

And here were forests ancient as the hills enfolding sunny spots of greenery.

Coleridge

The deserts of Iran and Mesopotamia (Iraq) tell the story, forest of gradual soil exhaustion and then of more rapid erosion. The mighty empires of Babylon, of Syria, Persia (Iran) and Carthage were destroyed by the advance of flood and deserts caused by the clearing of forests for agricultural land and by having their rivers drawn upon for additional supplies of water for irrigation. Their pristine fertility simply ran down under agriculture work out after long use.

The Tragedy of China

"When forests go, the waters go, the fish and game go, crops go, herds and flocks go, fertility departs. Thus the age old plantons appear, stealthily one after another – Flood, Drought, Fire, Famine and Pestilence."

B. Baker cites the example of yellow river which should be kept in mind while planning national projects in our country.

The yellow river and the yellow sea are named from the colour eroded soil which has been lost to the winterland. So great has been the erosion that often the beds of the rivers are raised like that of the Mississippi, until they are higher than the surrounding country sides. The water tear down the hill sides in ever increasing torrents and in China create the most disastrous floods that the world has seen. It has been estimated that the yellow river alone transports an annual load of 2500 million tonnes of soil. There are other region where rapid erosion is taking place, but the yellow river has been described as the outstanding and eternal symbol of the mortality of civilization.

But China look up massive soil conservation and afforestation works to counter the destructive processes and the catastrophe is now well under control

The Miracle of Japan

"Trees for fruitage and fine and shade. Trees for the cunning builders trade. He made them of very grain and girth for use of man in the Garden of Earth."

Bliss Carmen

"The Japanese are a people who have an inherent love for trees, and it has helped them in their struggle to save the land surface of their islands, which but for the well-tended forests, might have eroded away and suffered the same fate as Atlantis and the Continent of Mu. It is a miracle, indeed that these islands have survived so many volcanic upheavels and that their forests have contrived to prevent catastrophic erosion which would have inevitably doomed them to destruction."

The Miracle Island

By the water of Babylon we sat down and wept.

Psaln 137 : 1

To the traveller the islands of the Pacific are steeped in romance and they never fail to cast a mysterious spell over the mind of the voyager. The very mention of their names is as music : Hawaii, Thiti, Rarotonga, Bali, Java – these are as notes in an entrancing symphony. Many travel stories have been written about them; many are the romantic pages.

For the healing of the Nations

And the leaves of the tree were for the healing of the nations.

Rev 22:2

The health and economic security of the human race depend on how well the forests of the world are managed. All the countries of the world are suffering the penalty resulting from man's neglect to plant where he has reaped. He has destroyed the gifts of a generous creator without realising that they were a trust to be handed over on to future generations. The earth's green covering is Nature's Capital.

The forest is the mother of the rivers and are Nature's most important means of regulating and maintaining the flow and quality of water. We believe that the great task of conserving and replenishing the forests of the world and reclaiming the deserts and waste places by tree planting requires the concerted action of every country. We believe that in order to save humanity we must save men, and in order to save the forests we must save the trees.

World afforestation is necessary because it is the most constructive and peaceful enterprise.

Trees or Famine

The tree of the field is man's life.

Deut. 20:19

Trees are one of Nature's most efficient weapon of soil defence. Conservation farmer use them to tie down steep hill sides, to check the growth of big gullies, to stabilise unstable stream banks, and to screen off cultivated field from harmful winds.

The story of forests of India is an endless one, and it is impossible to tell in a small treatise of 300 pages.

Qualities and Functions of Forests
— Multifaceted Roles and Diverse Qualities of Forests

*F*orest play a very important role in maintaining ecological balance of the country.

Forest is a community of trees, shrubs, herbs, climbers along with various flora and fauna, which together creates an environment of its own. It consists of various types of plants, bearing flowers of various colours and smells, fruits of various shapes and sizes.

Foresters are intimately involved in protection, conservation and scientific management while performing their duties and they enjoy the enchanting environmental diversities and derives immense pleasures out of it.

In this chapter forest influences, interrelationship between forests, man and animals, man and wildlife, qualities and conception about forests have been discussed in nutshell. A number of charts have been presented for the readers to have overall ideas about these issues at a glance.

Qualities and Functions of Forests

– Multifaceted Roles and Diverse Qualities of Forests

FOREST INFLUENCES

The ill-effects of accumulated carbon dioxide can be minimised by vegetation which absorbs the gas by the action of photosynthesis. A well stocked ecosystem exercise considerable beneficial effects on the human environment by moderating the climate, maintaining the soil mantle, regulating the water supplies, purifying the air and helping in absorbing noise.

The most important environmental factors are micro climate, soil characteristics, moisture availability and action of animal and insects. Microclimate is governed by solar radiation, rainfall, wind, humidity and temperature of the air and the soil. The humidity is greater inside the forests. The soil temperature is influenced by the forests, which in turn affects the biological activity in the top layers of the soil. Forest stem the wind velocity. Extensive forests also condense low clouds and this to some extent increase precipitation.

In an open canopy forest there is little humus on the forest floors and therefore, rain strikes the almost bare soil with full fury and brings about soil erosion that leads to denudation.

Roots of trees penetrate deep into the soil and this bind it together. When the roots die they act as capillaries for water to flow downward. When the forest cover is destroyed, the protective leaf litter and humus are washed away, thus exposing the soil. In the rainy seasons the soil pores get clogged and percolation is inhibited with the result that water rushes in torrents carrying soil with it, thereby exposing the underlying rock. The Madan Mahal Hills of Jabalpur was covered with thick forests four centuries ago and Rani Durgavati used to hunt tigers in it (K. P. Sagreiya "Forests and Forestry") has now been denuded.

To form a comprehensive and realistic idea of multifaceted qualities of the forests, a chart has been drawn which will speak for itself.

A VISIONARY FORESTER
Dr. Ribbentrop
(An assessment made by him 150 years back)

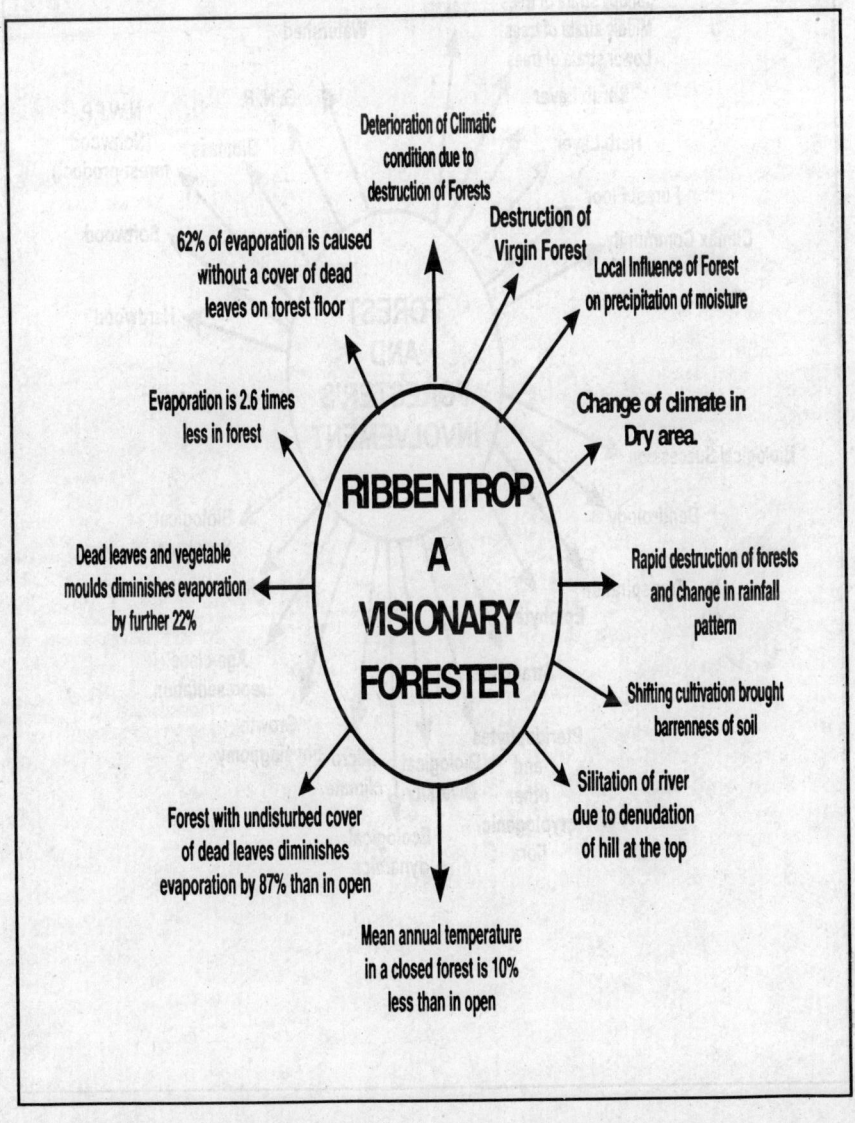

Deterioration of Climatic condition due to destruction of Forests

Destruction of Virgin Forest

62% of evaporation is caused without a cover of dead leaves on forest floor

Local Influence of Forest on precipitation of moisture

Evaporation is 2.6 times less in forest

Change of climate in Dry area.

RIBBENTROP A VISIONARY FORESTER

Dead leaves and vegetable moulds diminishes evaporation by further 22%

Rapid destruction of forests and change in rainfall pattern

Shifting cultivation brought barrenness of soil

Forest with undisturbed cover of dead leaves diminishes evaporation by 87% than in open

Silitation of river due to denudation of hill at the top

Mean annual temperature in a closed forest is 10% less than in open

A FEW FACETS AND FUNCTIONS OF FORESTS

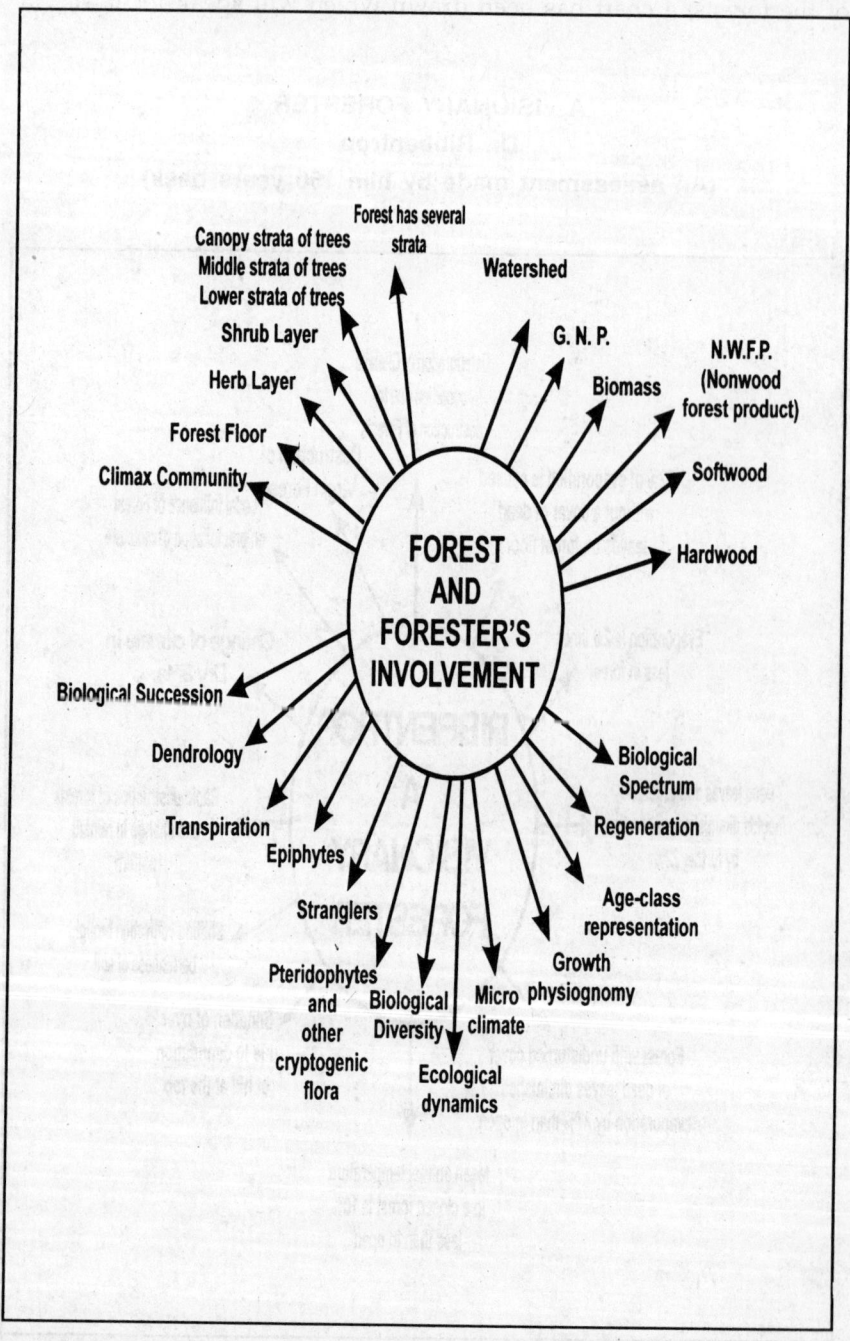

A WELL-MANAGED FOREST
CREATES AND CONTRIBUTES
TO DEVELOPMENT AND CONSERVATION

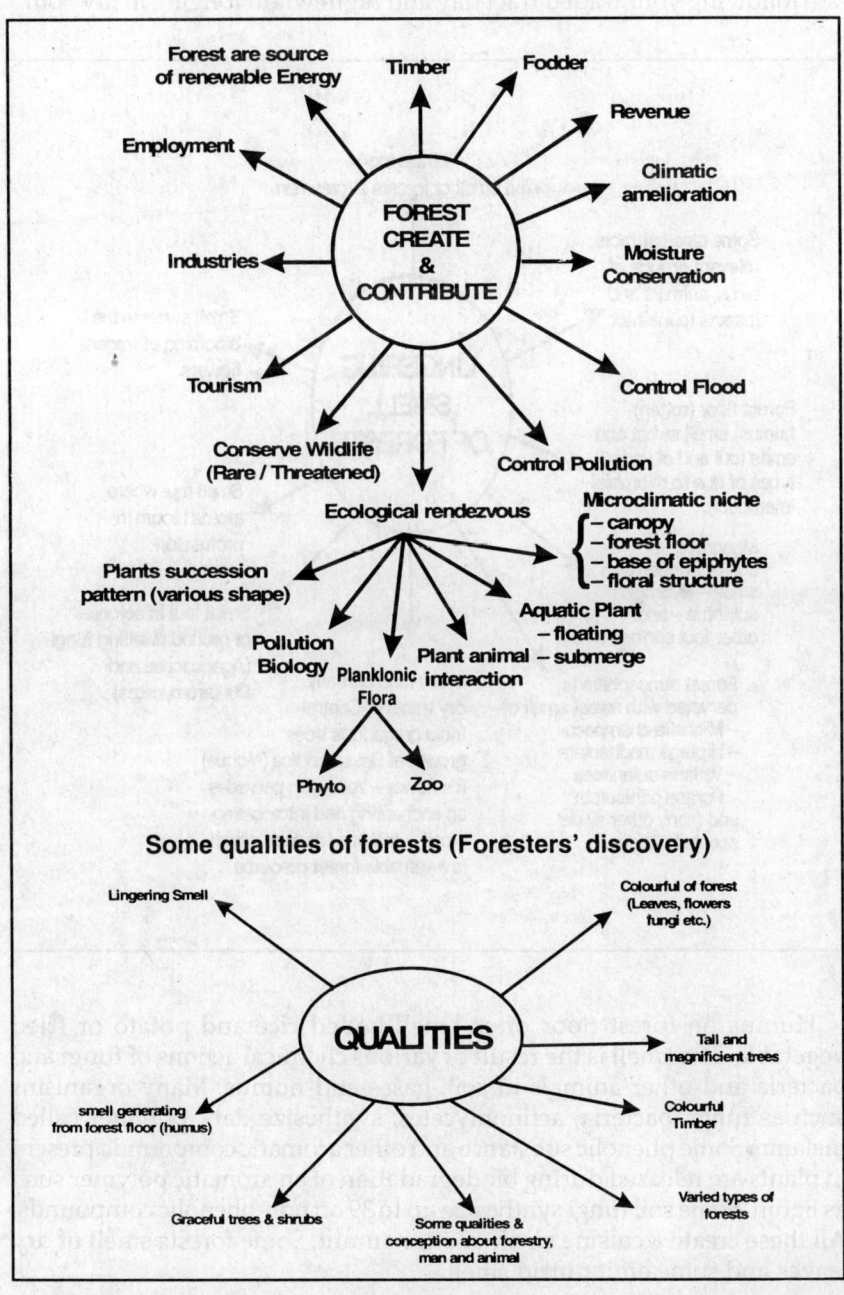

FOREST CREATE & CONTRIBUTE

- Forest are source of renewable Energy
- Timber
- Fodder
- Revenue
- Climatic amelioration
- Moisture Conservation
- Control Flood
- Control Pollution
- Employment
- Industries
- Tourism
- Conserve Wildlife (Rare / Threatened)

Ecological rendezvous

- Microclimatic niche
  - canopy
  - forest floor
  - base of epiphytes
  - floral structure
- Aquatic Plant
  - floating
  - submerge
- Plant animal interaction
- Planktonic Flora
  - Phyto
  - Zoo
- Pollution Biology
- Plants succession pattern (various shape)

Some qualities of forests (Foresters' discovery)

QUALITIES

- Lingering Smell
- Colourful of forest (Leaves, flowers fungi etc.)
- Tall and magnificient trees
- Colourful Timber
- Varied types of forests
- Some qualities & conception about forestry, man and animal
- Graceful trees & shrubs
- smell generating from forest floor (humus)

Smell of Forests (each forest has it's distinct smell)

"Lingering wild perfumed smell your presence is ever leading me on. I am following your traded tract day and night wight longing in my Soul."

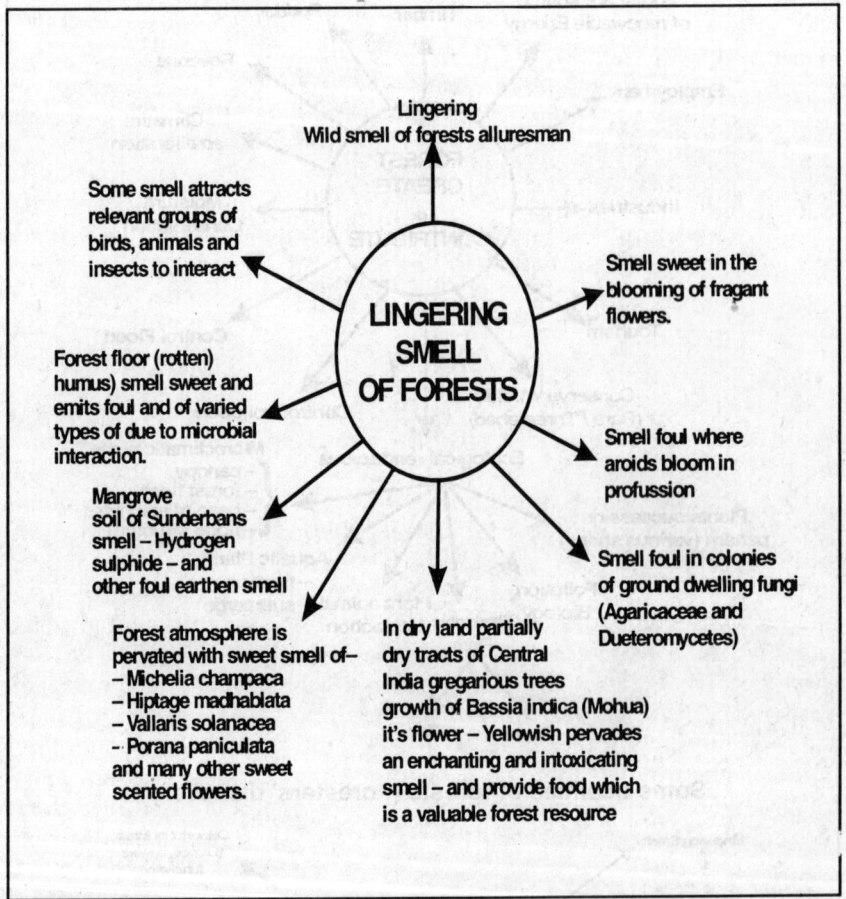

Humus on forest floor often smell boiled rice and potato or fried vegetable. The smell is the result of various chemical actions of fungi and bacteria and other animals in soil, gases and humus. Many organisms such as fungi, bacteria, actinomycetes, synthesize dark polymer called melanin. Some phenolic substance and other aromatic compounds present in plants are released during biodegradation of an aromatic polymer such as lignin. Some soil fungi synthesize up to 39 on non-phenolic compounds. All these create a cuisine smell of a restaurant. Some forests smell of dry leaves and some emit putrid smell.

COLOUR IN FORESTS

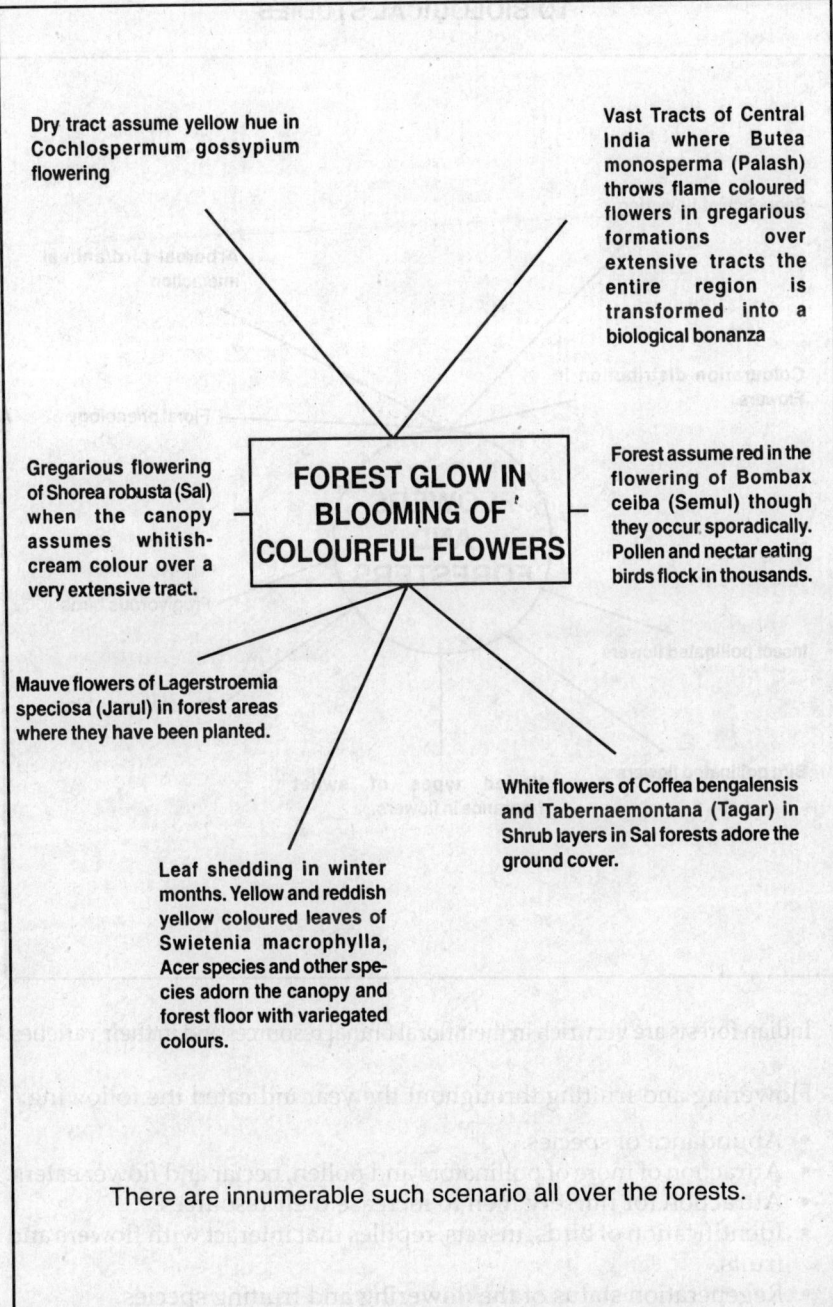

Dry tract assume yellow hue in Cochlospermum gossypium flowering

Vast Tracts of Central India where Butea monosperma (Palash) throws flame coloured flowers in gregarious formations over extensive tracts the entire region is transformed into a biological bonanza

Gregarious flowering of Shorea robusta (Sal) when the canopy assumes whitish-cream colour over a very extensive tract.

FOREST GLOW IN BLOOMING OF COLOURFUL FLOWERS

Forest assume red in the flowering of Bombax ceiba (Semul) though they occur sporadically. Pollen and nectar eating birds flock in thousands.

Mauve flowers of Lagerstroemia speciosa (Jarul) in forest areas where they have been planted.

White flowers of Coffea bengalensis and Tabernaemontana (Tagar) in Shrub layers in Sal forests adore the ground cover.

Leaf shedding in winter months. Yellow and reddish yellow coloured leaves of Swietenia macrophylla, Acer species and other species adorn the canopy and forest floor with variegated colours.

There are innumerable such scenario all over the forests.

**FLOWERS ARE GIFT OF NATURE GOD AND FORESTER'S
INVALUABLE RESOURCES CONTRIBUTION
TO BIOLOGICAL STUDIES**

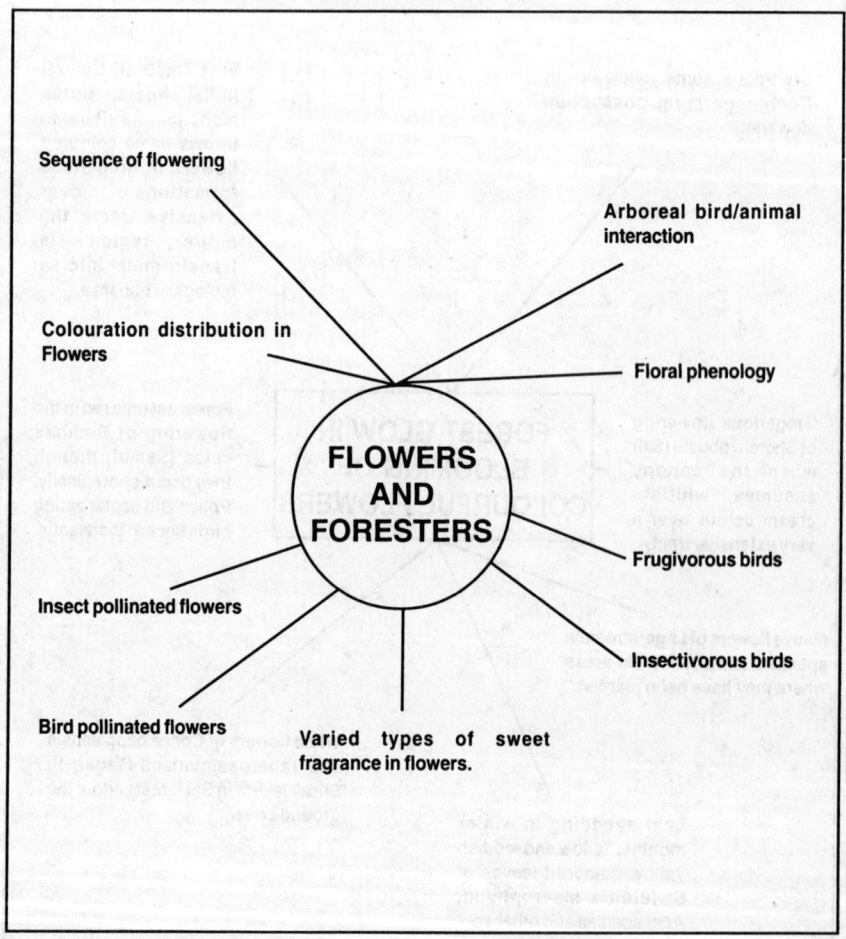

Indian forests are very rich in their floral faunal resources and in their varieties.

Flowering and fruiting throughout the year indicated the following:

- Abundance of species.
- Attraction of more of pollinators and pollen, nectar and flower eaters.
- Attraction for nursery men to increase their resources.
- Identification of birds, insects, reptiles that interact with flowers and fruits.
- Regeneration status of the flowering and fruiting species.

TALL AND MAGNIFICENT TREES OF INDIAN FOREST

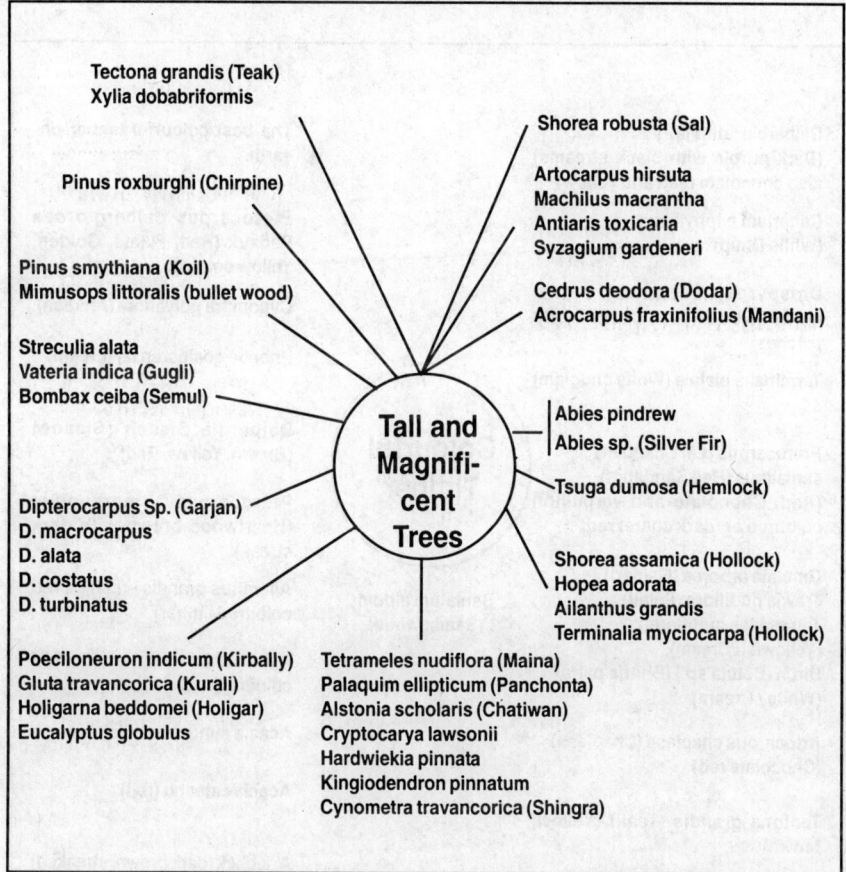

Tectona grandis (Teak)
Xylia dobabriformis

Pinus roxburghi (Chirpine)

Pinus smythiana (Koil)
Mimusops littoralis (bullet wood)

Streculia alata
Vateria indica (Gugli)
Bombax ceiba (Semul)

Dipterocarpus Sp. (Garjan)
D. macrocarpus
D. alata
D. costatus
D. turbinatus

Poeciloneuron indicum (Kirbally)
Gluta travancorica (Kurali)
Holigarna beddomei (Holigar)
Eucalyptus globulus

Tall and Magnificent Trees

Shorea robusta (Sal)

Artocarpus hirsuta
Machilus macrantha
Antiaris toxicaria
Syzagium gardeneri

Cedrus deodora (Dodar)
Acrocarpus fraxinifolius (Mandani)

Abies pindrew
Abies sp. (Silver Fir)

Tsuga dumosa (Hemlock)

Shorea assamica (Hollock)
Hopea adorata
Ailanthus grandis
Terminalia myciocarpa (Hollock)

Tetrameles nudiflora (Maina)
Palaquium ellipticum (Panchonta)
Alstonia scholaris (Chatiwan)
Cryptocarya lawsonii
Hardwiekia pinnata
Kingiodendron pinnatum
Cynometra travancorica (Shingra)

These mighty and magnificent tree species occur in India's 16 different agro-climatic zones, 10 broad phytogeographical divisions, more than 25 biotic provinces, 200 forest types and about 450 biomes (habitat of specific species).

Recorded Height

Eucalyptus	–	65.0 m
Chirpine	–	65.5 m
Deodar	–	64.9 m
Padauk	–	51.5 m
Sal	–	51.2 m
Teak	–	43.0 m
World Tallest Redwood	–	110.0 m

COLOURFUL (decorated)
TIMBER (a few examples)

Dalbergia latifolia
(Dark purple with black streams)
also chocolate read and yellow)

Canarium euphyllum
(White Dhup)

Diospyros marmarata – the
marblewood (Black and white
patches)

Terminalia bialata (White chuglam)

Prerocarpus (Lal Chandan)
santalinus (Red Sandaurs)
(Red, Chocolate and vermillion
coloured or (dark charet red)

Gmelina arborea (Gamari)
Trewia nudiflora (Petali)
Tetrameles nudiflora
(Yellowish/Cream)
Birch (Betula sp.) (Bhurja patra)
(White / Cream)

Artocarpus chaplasa (Chaplash)
(Chocolate red)

Tectona grandis (Teak) (Yellow,
brown)

Disopyros ebony (ebony) (Black)

Colourful Timber

Sanlatum album
(Sandalwood)

Syszyguim gardneri (dark
reddish brown with yellow
patches)

The best colourful timber on
earth.

Pterocarpus dalbergiordes
Padauk (Red, Flame, Golden,
Yellow colour)

Cynomitra polyandra (reddish)

Phoebe goalparensis (cream)

Dalbergia Sissoo (Sissoo)
(Brown, Yellow, Red)

Pterocarpus marsupium
(Heartwood brown with dark
streak)

Ailanthus grandis – Lali (a red
coloured timber)

Aphanamixis polystachia (Red
coloured)

Acacia sundra (reddioh brown)

Acacia catechu (red)

A. lebbek (dark brown, streaked)

A. odoratissima (dark brown)

India has more than 1200 tree species, each one with it's distinct colour. The author considers Andaman Padauk is the best colourful and quality-wise top class in the whole of the world; the marblewood has also an attractive black and white combination. Teak has universal acceptance.

GRACEFUL PLANTS

Some of the very many species are listed below:

Astonia scholaris	Acer campbellii	Azadirachta india
Dillenia indica	Pongamia pinnata	Saraca asok
Ficus benjamina	Exbucklandia populnea	Brownea hybrida
F. clavata		Amherstria novilis
Betula utilis		Callistemon viminalis
Polyalthia pendula	**Graceful Plants**	Salix babylonica
Magnolia campbellii		Aralia foliolosa
Mesua ferrea	Terminalia catapa	Putranjiva roxburghii
Memecylon edule	Plumeria rubra	Celtis tetrandra
Filicium decipiens	Daphniphyllum himalayense	Averrhoa corambola
Saurauya roxburghii	Erithrina glauca	Thespesia populnea
Ilex godajam	Fiscus elastica	
Tecoma undulata	Cassia fistula	
Ficus incisa	Ternstroemia gymnanthera	
Ficus hookerii		
Ficus elastica	Camellia kissii	
Caryota urens		
Phoenix rupicola	Cedrus deodara	
Pandanus fasciculatus	Cupressus sp.	
Coroupita guinessis	Pinus sp.	
Cresentia cajute	Abies webbeiana	

VARIED TYPES OF FORESTS AND MANAGEMENT SYSTEMS

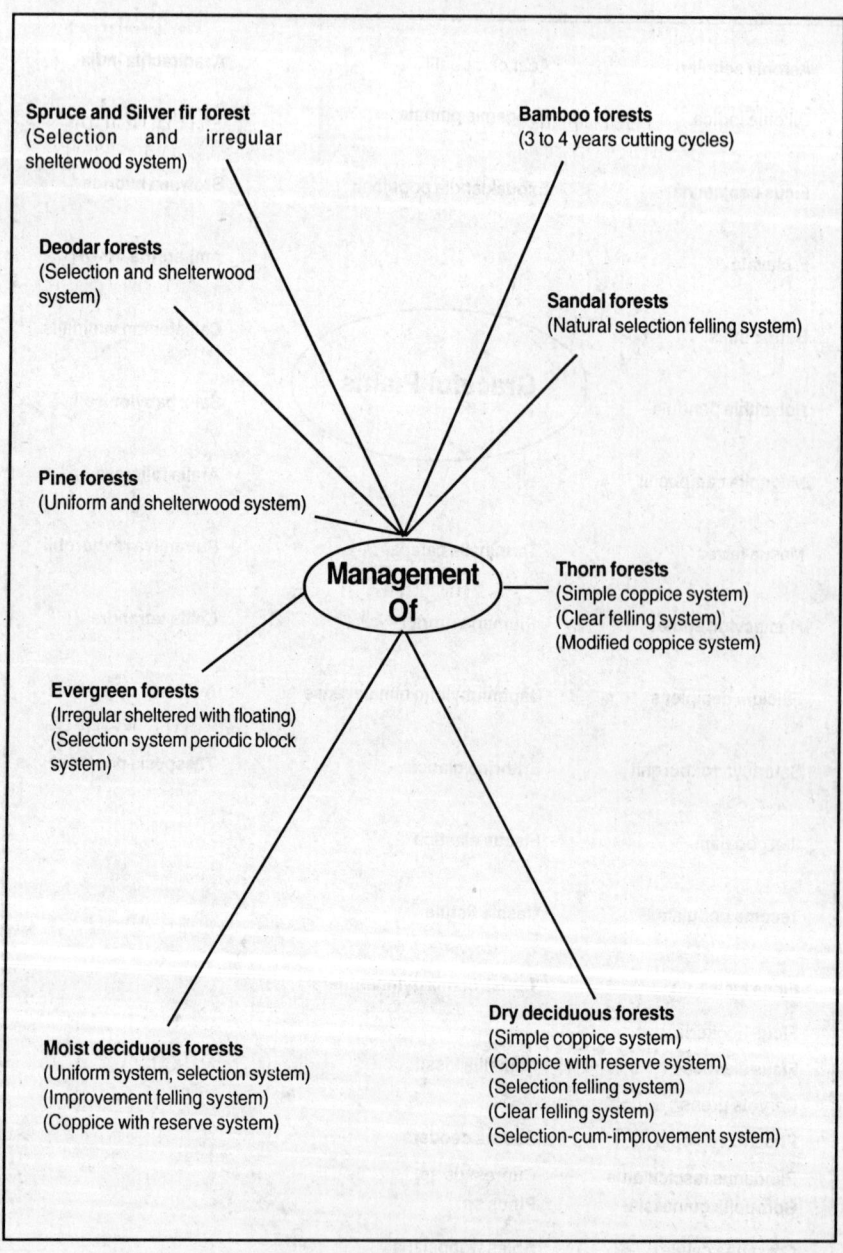

Spruce and Silver fir forest
(Selection and irregular shelterwood system)

Bamboo forests
(3 to 4 years cutting cycles)

Deodar forests
(Selection and shelterwood system)

Sandal forests
(Natural selection felling system)

Pine forests
(Uniform and shelterwood system)

Management Of

Thorn forests
(Simple coppice system)
(Clear felling system)
(Modified coppice system)

Evergreen forests
(Irregular sheltered with floating)
(Selection system periodic block system)

Moist deciduous forests
(Uniform system, selection system)
(Improvement felling system)
(Coppice with reserve system)

Dry deciduous forests
(Simple coppice system)
(Coppice with reserve system)
(Selection felling system)
(Clear felling system)
(Selection-cum-improvement system)

N.B. : Various complicated management issues are not been discussed

Some Qualities and Conception about Forests

"When the forests go, the water go, the fish and game go, crops go, herds and flocks go, fertility departs. Then the age old phantoms appear, stealthily one after another – Flood, Drought, Fire, Famine, Pestilence."

[**Robert Chamber**]

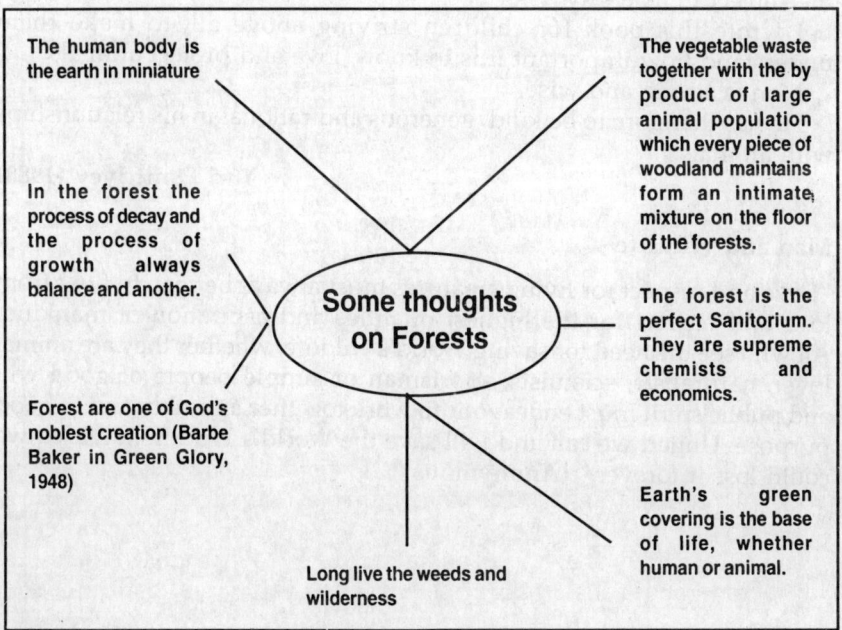

The human body is the earth in miniature

In the forest the process of decay and the process of growth always balance and another.

Forest are one of God's noblest creation (Barbe Baker in Green Glory, 1948)

Some thoughts on Forests

The vegetable waste together with the by product of large animal population which every piece of woodland maintains form an intimate mixture on the floor of the forests.

The forest is the perfect Sanitorium. They are supreme chemists and economics.

Earth's green covering is the base of life, whether human or animal.

Long live the weeds and wilderness

"There is ruthless rhythm in the intensity of growth where everything germinates and begins to thrust its way upwards in bloodless conflict for place in the Sun."

[*Source:* **Sir Albert Howard 1938 in his paper "The Floor on the Forests"**]

Forests, Man and Animals

"Man's life on earth is bound up fast with animals, birds and fish, insects and beasts, octopi and worms.

But over the millennia man's relationships with animals have undergone many changes. Animals supplied man with food and clothes, inspired him with fear and gave him joy, originated customs and beliefs that sometimes influences the entire mode of life of the given society, they have been enemies, friends and tutors.

God – animals were replaced by worker animals, wild animals were replaced by domestic ones as the main source of meal. The importance of some animals was enhanced and of other diminished. And, naturally, throughout Man's history he has brought influence to bear in the animal world – directly or indirectly consciously or unconsciously.

One book is not really enough to tell about many different relationships between man and animals. Nor have I tried to embrace the subject in its entity.

I wrote this book for children striving above all, to make them understand how important it is to know, love and protect animals.

Man is strong and wise.

He has therefore to be kind, generous and rational in his relationships with animals."

Yuri Dmitriyev (1983)

Man and Wildlife

"Love and respect for living creatures must always be seen to flow from love and respect for the highest qualities and aspiration of mankind. All who see the need for saving World's wildlife, whether they are animal lover, naturalists, scientists, sportsman or simple people of good will and public spirit must endeavour to work together amicably and to good purpose. United we can and will save the World's Wildlife, divided we could loss it forever." **[Anonymous]**

MELODY IN FORESTS – BIRDS OF INDIA

Grey hornbill
(a loud cackling k-k-k Kae and variety of squealing and chattering conversational tones)

Brown head Stork-billed Kingfisher
(a rameous explosive chattering laugh Ke-ke-ke)

Whitebreasted Kingfisher
(a loud cackling frequently repeated scream)

Treepie-Melodious
(Kokila/bob-o-link)

Grey Tit-whistling
(Whee-Chichi)

Grey Tit-whistling
(Whee-Chichi)

Yellow Chicked Tit
Whistling (Cheewit-pretty-cheewit)

Velvet fronted Nuthatch
(Loud cheeping whistle)

Common Babbler
(Series of pleasant trilling whistle)

Spotted Babbler
(3-4 rich paintive whistling note)

Shyama
(Loud clear melodious thrush like song)

Golden Oriole
(A harsh cheah and dear fluty whistles like peelolo)

Hill myna
(An accomplished mimic and talker)

Birds and Call of the Forest

Spotted Owlet
(A large variety of harsh chattering, squabbling chuckling note)

Collared Scops owl
(A soft interrogative repeated monotonously every 2-3 seconds over long stretches of time)

Indian great Horned Owl
(A deep solemn resounding bu-bo)

Brown Fish Owl
(A deep hallow meaning Boom-o-boom with peculiar eerie ventriloquistic utterances)

Barn or Screech Owl
(A mixture of harsh discordant screams and weird shoring and hissing note)

Common Indian Nightger
(Chuk-chukchuk)

Hoopoe
(Soft musical penetrating hoo-po-po)

Pied Kingfisher
(A sharp cherry, Chirruk, Churruk)

Common Swallow
(A pleasant twillaing)

Copper Smith
(Loud monotonous ringing, tuk tuk repeated every second)

It is a futile exercise to give impression in words of language about the melodious note of birds. One can hear, feel and appreciate the song. Only a few of 1200 species of Indian birds are mentioned above. It is a feeling that people should desire to be among many of these 1200 species at specific sites.

A Well-Protected Forest Creates Environment
— Forester's Contribution

A forest creates its own environment, when it is well protected. It invites visitors, acts as shelter for animals and birds, generates myriads of sounds, forms the cloud, gives birth to rivers and streams, in addition to its numerous direct and indirect benefits.

In this chapter all the above issues along with the role of forest in respect of inviting visitors, providing scope for scientific studies and its other educational and entertainment values have been dealt in short.

A Well-Protected Forest Creates Environment

— Forester's Contribution

Forests are eternal source of endless natural qualities. These qualities when enumerated, are sure to involve a large space in compilation. But

A FEW EXAMPLES OF SUCH QUALITIES

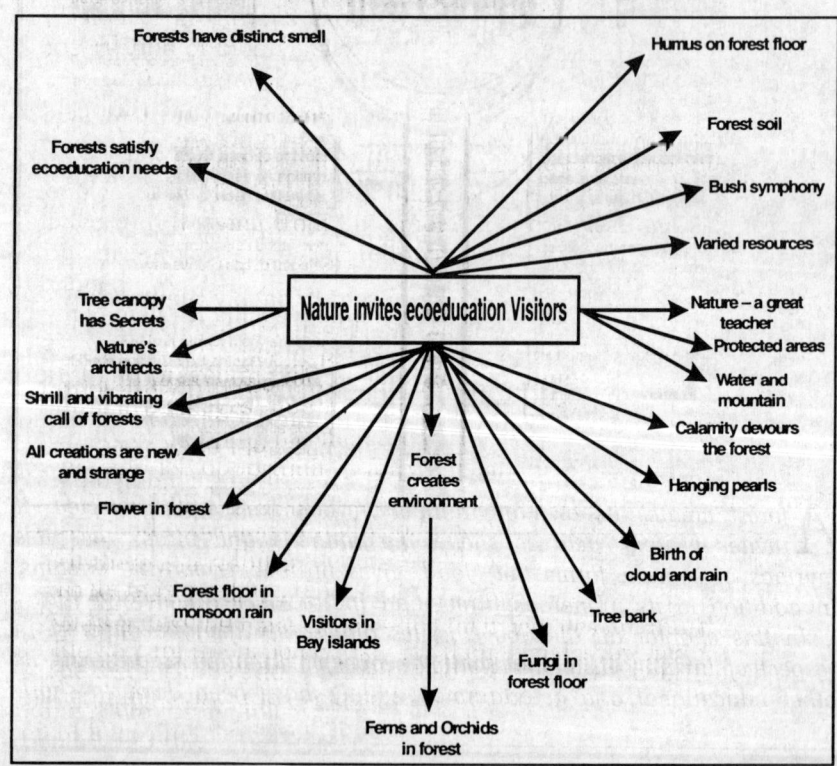

some qualities cannot be ignored and the author has picked up a few interesting qualities to impress upon the readers how diverse, environment friendly and educative such qualities are. In choosing these few qualities the author has selected some phenomena that the young school and college students, the visitors and the tourists in the forest would like to know.

Foresters have discovered and preserved many qualities of nature and environment. The author has picked up a few such qualities in several chapters. The forests have many such multifaceted qualities for posterity to uncover, study and enjoy.

Secrets of Forests in Tree Canopy

Flowers bloom in 'Rain forest' once in 5 to 6 years, when entire forest burst into flower at a time (an extraordinary synchronized event). The pattern of life in Sal or Gurjan forests represent a panicle of arboreal evolution in Indian panorama.

The canopy is arguably the most species – rich environment of the plant, and hence is termed the "last biotic frontiers" (Erwen 1988). Canopy dwellers are snakes, anurans and lizards, etc. Forest canopies are structurally complex and ecologically diverse.

Forest canopy is the primary site of gas exchange between atmosphere and vegetation. Forest canopy is a three dimensional subsystem of the forest itself. There are many species of epiphytes-vascular and non-vascular, insect, ants, etc. and all these form a biological unit and a unique food chain. Many of these species have evolved themselves over many years — a unique feature for rain forest.

Hanging Pearls

Rain drops or dew accumulate around the glue spots in innumerable nets on spider's webs made from silk secreted through apertures on the spinnerets; they glow pearl like in reflected and diffused sunlight under forest canopy which are things of joy and enjoyment.

These are common sights in the rain and wet mixed forests of India.

Shrill and Vibrating Call of Forest

Myriad sounds generate inside the forests which originate from animal, animal-animal interaction or from wind/forest plant interaction.

The most common is the call of Cicadas which pierces the silence of forest by shrill and vibrating resounding call. It creates horror when several such insects start a music in a chorus until the whole forests tremble with noise. They have long membraneous wings, frequently brilliantly coloured as they camouflage on the bark. The sound may be compared is that of a high speed circular saw or a steam whistle and is certainly not equalled by no other insects.

A FEW ARBOREAL BONANZA ARE AS FOLLOWS

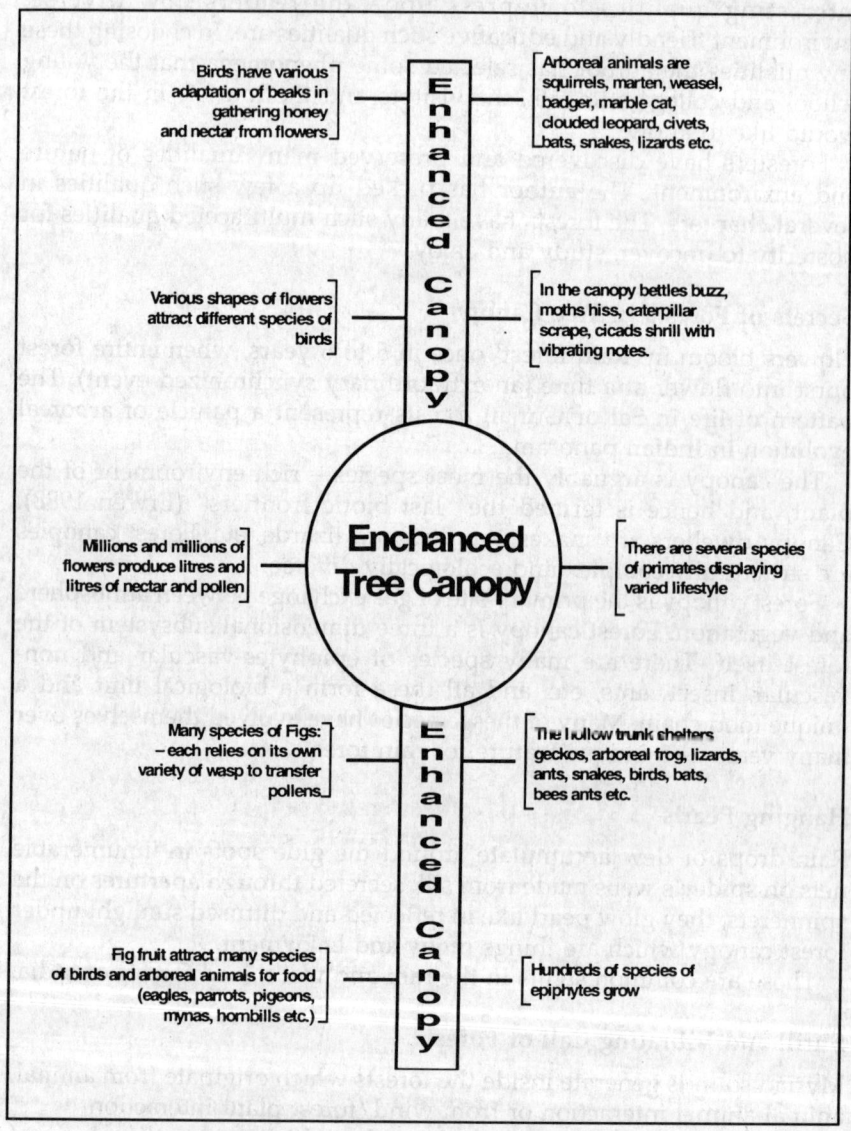

Birds have various adaptation of beaks in gathering honey and nectar from flowers

Arboreal animals are squirrels, marten, weasel, badger, marble cat, clouded leopard, civets, bats, snakes, lizards etc.

Various shapes of flowers attract different species of birds

In the canopy bettles buzz, moths hiss, caterpillar scrape, cicads shrill with vibrating notes.

Enchanced Tree Canopy

Millions and millions of flowers produce litres and litres of nectar and pollen

There are several species of primates displaying varied lifestyle

Many species of Figs: – each relies on its own variety of wasp to transfer pollens

The hollow trunk shelters geckos, arboreal frog, lizards, ants, snakes, birds, bats, bees ants etc.

Fig fruit attract many species of birds and arboreal animals for food (eagles, parrots, pigeons, mynas, hornbills etc.)

Hundreds of species of epiphytes grow

Enhanced Canopy

But there is a respite.
An adage records –
"Blessed are the male cicads – for their females are mute."

The author may add – blessed also are the forest goer who have to spend long time in the forests.

Birth of Cloud and Rain – An Example from Amazon Forests

That forests create rains is best known from various studies made in Amazon rain forest in South America. One finds an endless sea of vegetation through the centre of which runs world's largest river – Amazon. It drains more than 6 million sq.km and has more than 10,000 tributaries. It carries two-thirds of the worlds free-fresh water. The mouth of the river is 240 km across and it is tidal up to 600 km inland. From evaporation and transpiration swirling mists accumulated as thick blankets and their cumulative actions from cloud and eventually rain. Such phenomena are also visible in rain and even wetmixed or moist deciduous forests of India to a lesser extent; mist and cloud formations are commonly sighted in such forests.

Water and Mountain

These are new creations which the budding school, college students should note the wild animals, the fish, the birds, the reptiles, the plants, the bees, the flowers – all are new, beautiful and grotesque and will be the sustained sources of attraction of the visitors.

Nothing can surpass the circumbient winding and romantic banks of a narrow arms of creek; the over head most grotesque foliage of rain forest trees, their perfumed shadows, picturesque mangrove of the bays and rivers are all clothed with thick forests over extensive areas.

Visitors in Andaman and Nicobar Islands

The infrastructure of touring facilities here may not offer them a dive by day, dance by night, and enjoy and exotic tropical cocktail or the warm sands of the beaches as frothy little waves lap at your feet. But there are equally attractive spots as in Phuket of Thailand, one of world's best scubadiving destinations, which is not far from our Islands. The sea here offers warm water, excellent underwater visibility, myriad varieties of marine life, opportunity of swimming with whale, sharks and manta rays. In Phuket Kings regatta hundreds of worlds keenest ocean sailors assemble for some spectacular racing. The clear water of the Andaman sea are also ideal for gamefish, and the region is in the migratory paths of classic fighting fish. Phuket even hosts its own surfing competition. The Phuket quicksilver surfing contest at Kata beach. Here, if you feel like catching some waves, check-out the local beaches between June and October swimming, snorkelling, dingy and catamaran sailing wind surfing, jet-skiing, wind surfing jet skis, water skiing and para sailing can be enjoyed here. Andaman Island can be developed to serve such facilities in future but Phuket cannot educate the tourists as the Andaman islands do.

Forest Floor in Monsoon Rain

Moisture laden wind transform into welcome rain when myriads of insects, frogs toads and invertebrates emerge and the forest reverberates with the call of birds and animals.

Seed germinates, under ground roots send flowering stalk and entire forest floor dances colourful dance.

Most interesting is the sighting of frogs welcoming the visitors with an orchestra of mating calls.

The monsoon is a season of plenty and creativity. Streams and brooks are born, saplings and seedlings nourish, seeds sprout, flowers bloom, snails crawl and gorge on new buds, butterflies tango with multihued flowers. Innumerable such activities can be sighted.

Nature is a Great Teacher

A visitor of forests will observe a multi-storeyed forest of several tiers, each tier having a distinct structure and separate role to play. There are birds and arboreal animals at each tier and they have specific architectural technique in building their nests and continue life processes. Various woody climbers (lianas) have specific twisted boles; epiphytes like bird nest ferns (Asplenium) form special niche for several arboreal animals. There are many such examples.

"All Creations are New and Strange" (Anonymous)

The forest visitor will feel the magnificence of nature in all its denizens when he enters into the forest. Most of the spots within the forest are new heaven and a new creation. In the hills tiers beyond tiers, height above height, the great wooded ranges go rolling away northwards till on the lofty Himalayan skyline they are crowned with a gleam of everlasting snow. In plains wooded tract runs kilometres after kilometres across streams, rivers, creeks, tea estates, and bays that reveal nature's bounty.

A visitor to the forest may note to the Chinese proverb which read as follows.

- I reap, I forget
  I see, and I remember
  I do, and I understand

- Ten thousand words
  are not worth one seeing

- You give a man a fish
        You feed him for a day.
  You teach them to fish
        You feed him everyday.

The visitors therefore are to keep their eyes open while trekking through of the forests.

They may know an adage from Confucius which reads: "Wise men find pleasure in water, The virtuous in mountains"

The visitor of various forest areas are both virtuous and wise as they will have to learn from both water and mountain wilderness and should read the voices of the hills and rivers.

The visitors are virtuous to have come to visit the various forests for pleasure and study. They feel oneness with the forests and entire atmosphere as soon as they touch the forest floor, breathe in pure and fragrant air and feel cuisine smell of the forest. They are mesmerized at the sight of multilayered forests land rich wildlife species; the enhanced forest canopy and its varied and rare resources transform them and from a mere visitor to researchers. Some of these alterations have been enumerated in a few paragraphs.

Forests Satisfy a Wide Range of Ecoeducation Requirements

Sal, Gurjan, Sishum, Pine, Fir and other trees, the murmuring streams and the serene calmness of the forests bring joy and relief to forest lovers. The forests satisfy their various intellectual and esthetic requirements of people.

A Chinese proverb says –
"The legs of a crane-fly is long;
The legs of a lady bird is short,
Why worry?"

The forests are store house of vast species of flora, fauna on varied soil inhabited by millions of microscopic invertebrate. All these create various complex environmental facets found nowhere else on this earth.

Tree Bark – An Indicator for Identification of Species

They are of variegated coloured, textured and forms; when blazed they give sweet, sour and wild smell, tastes various and they ooze white, red and yellow juice. Foresters identify exact species after examining all these qualities.

The pattern and colour of barks of various trees give each species its identification character.

Calamity Devours the Forest

Forest fire creates horrors, damages timber wealth, young regeneration, ground dwelling wildlife, soil fauna and ground flora. Fire sometimes proceeds at an appalling speed, from peak to peak, and blows wind to a great force.

Flood wipes out large chunks of forest, deposit sand and silt on forest floor killing trees large areas. Other calamities are due to drought and diseases all these kill large number of trees.

Numerous instances may be cited about such calamities that have damaged vast tracts of forests and beautiful country side.

Fungi on Forest Floor

Fungi display a bewildering variety of forms and colour on forest floor. Most striking and attractive scenario is the occurrence of *Polystictus sanguineus* an orange coloured Polyporaceae that beautifies the forest floor; others that grow on humus are Agaricus and Phallus, besides Hexagonia, Lentinus, Hydnum, Ganoderma, Irpex, Polyporus, etc. on rotten wood. Daedelia flavida, a polypore, is white in colour that decorate forest floor.

Flowers in the Forest

Red, pink, purple, white and yellow, a myriad hues of flowers greet the visitors with their colour, colour combination, colourful butterflies, hovering over them. Breathe taking descriptions have been given by the explorers in late eighteenth, nineteenth and twentieth centuries. There are many illustrated publications that the visitors may peruse.

Soil

Soil is natural resource; all our food comes from it directly or indirectly. This earth's uppermost layer provides organic as well as mineral water for the growth of plants. The organic matter and humus are derived from the decomposition of plants and animals, where as the inorganic or mineral water result from the sub-aerial denudation of rock. Over 90 percents of the average soil is inorganic. Mineral water is derived from original rock weathering. Soil is made up of substance existing in three states – solid, liquid and gaseous. It is not a lifeless residual layer, accumulated over a long period of years. It is a dynamic layer is which many complex chemical physical, and biological activities are going on continuously.

The liquid present in the soil is a complex solution capable of engaging in a multitude of important chemical reactions. Gases in soil are those liberated by biological activities and chemical reaction in soil.

Erosion of various types affected the forest floor as well as agricultural land. The water holding capacity of soil has declined considerably all over the country. There are vast tracts of eroded land all over the country.

Bush Symphony

In various forests visitors may hear a symphony of orchestra of the bush where varied types of musical, crackling and melodious sound emanates from inside the forests. These are due to wind action on branches and leaves of trees interaction of insects, birds and other myriads of forest denizens in various phases of their lives in different seasons.

Humus in Forest

Humus is like a blotting paper of the forest floor. As one gets down in his vehicle and steps on the forest floor, his feet get a valvety touch as they sink. This substratum is humus – which is dark brown soil mixed with decomposed plant and animal parts due to bacterial and fungal decomposition. This humus has ability to absorb water as they are soft and spongy. Numerous vertebrate and invertebrate species interact in this layer. Humus contains the main element of plant nutrition. They are sometimes also aromatic.

Humus (containing fulvic and humic acid) are non-living finally divided organic matter in soil derived from microbial decomposition of plants and animals substance; colour ranges from brown to black. (Carbon form 60%, nitrogen 6%, other sulphur, phosphorus, etc. – the role of various fungi, invertebrate earthworm, termites in humus formation is a complex subject).

Although humus is a mixture of numerous organic substance, two types of polymers, humic acid and polysaccharides from the major fractions. Humic acid is a complex polymer of hydroxyphenal, hydroxy benzoic acids and other aromatic structure with linked peptides, amino, sugar, fatty aids, microbial cell wall and protoplasmic fractions.

It takes hundreds of years for a thick humus layer to grow. Fire reduces such layers to ashes in no time.

The situation was not hopeful in most of the accessible forest areas in the past. It is worse now even in some inaccessible areas due to grazing and fire and other biotic impacts.

Ferns and Orchids

Fern flora of wide varieties of leaf form provide decorative materials in the towns and cities in various functions. Orchids are invaluable gift of nature. In Darjeeling, Sikkim Himalayas the orchids are very rich natural as crop; besides, the orchid flowers are widely cultivated in several nurseries under competent management. Ferns occur in great abundance on moist forest floor.

The Protected Areas

These are sacred place where time stands still and silence rules, where crime, war, pollution and endless development do not disturb nature. Such areas rejuvenate human spirit. These are like a well protected natural museum. Many such areas have been set apart by the foresters in varied forest types at different geographical and altitudinal ranges.

Wild Scope of Study on Forest Resources

Ecoeducation explorers will have innumerable facets to study the nature and forests concerning flowers, fruits, epiphytes, parasites, birds, insects, transpiration, respiration, etc. Within the canopy itself several microclimates are produced which turn the entire canopy layer into a bioecologically rich zone.

The colourful birds, butterflies and moths have captured human imagination like flowers which are nature's divine gift. Human ingenuity have exploited their brightness, kaleidoscope of colours and symmetry for hundreds of years. Butterflies are wonderful and beautiful creatures. People marvel at the assemblage of varieties of colours and patterns of their wings. Their complex life history from hatching of eggs to full fledged insect puzzles the science. Moth and butterfly species are part of the heritage of India. Most important is the dependence of larvae on specific species or families of plants for food in pupal stage. Depletion of such plant species from forest are affecting their future regeneration.

Nature's Architects

The role played by birds, spiders, termites, marmots and many other wasps, insects, birds and animals in building their place of abode show a high quality architecture. Nests, tunnels, burrows show their ingenious skill. Ants and bees build their colonies that provides separate chambers for various activities and function baffle the scientists. Termites are legendary builders (ant's saliva mixed with mud). The honey comb is a miracle of construction built primarily of secreted wax; a bee hive is no less complex or mind-boggling in its conception, design and function. A well protected forest offers all such varieties.

Nature Invites Ecoeducation Visitors

Nature has inexhaustible resources. The area covered by the forest opens up millions of facets of which bio-ecological studies are a part of this vast resources. Moreover the discussion is very much limited considering the limited scope to keep the size of the presentation to about 350 pages. The author had to chose a few attractive and conspicuous qualities of nature, especially of forests, to evoke interests among the budding school, college student nature lovers and ecoeducationists.

The author all through his service period saw young ones are concerned more after the thrill of seeing a wild animal and listing of shikar episodes than learning from vast natural laboratory or museum.

5

Conserving Biodiversity Resources
— Forester's Effort to Save the Threatened Wildlife

*T*he concept of conservation centres round the wise maintenance and utilization of the earth's resources, which should be managed in such a way that the resource is not exhausted. It concerns both biological and physical diversity which is to be maintained, besides the conservation of genetic materials. This is a recent concept but the foresters created many protected areas all over the country. Ecodevelopment stretch an ecosystem to provide more food and commodities for human development without causing a breakdown of natural system – a symbiosis of human kind and environment.

Biodiversity means the variability among living organism from all sources, inter alia, terrestrial, marine and other aquatic ecosystems and the ecological complexes of which they are part, this includes diversity within species, between species and of ecosystem.

Forester's effort are commendable in both these fields among many limitations.

5

Conserving Biodiversity Resources

— Forester's Effort to Save the Threatened Wildlife

India has a very extensive network of protected areas (National Parks, Sanctuaries, Tiger reserves, etc.) covering more than 1,50,000 sq.km. Most fascinating and prestigious protected areas are project Tiger areas, Keoladeo National Park, Kaziranga National Parks and many other such areas.

Yet, the animals are not fully secured inside these protected areas. Financial constraints, crippling shortage of adequate staff, poaching, insurgency, inadequate communication system, lack of judicial support are some deficiencies besides several others.

Some of these detrimental factors are responsible for various animals insecurity inside the protected areas. The export of tiger and leopard skins, bones, claws are rampant as may be witnessed from frequent illegal traders being held with booty by the customs, forest and police officials.

The forest covering 6,00,000 sq. km., are under control of Forest Department, a quarter area of it is under the umbrella of Protected areas. The very network that preserved the green cover and its denizens over many decades is now falling apart.

Sometime past about half a million mature sal trees were felled in Madhya Pradesh ostensibly to prevent spread of a insect epidemic. However, the clear cutting was halted by the Court. Similar situation is common in many states.

Even zoo animals have become vulnerable to poachers. The incidence of Hyderabad, Nandankanan are two such examples. Pilferage of bird, animal skins and various other parts of their bodies, even the live birds and animals are being detected all over the country. The official registers of the Govt. of India and Export-Import Trade Control offices testify to the type of prohibited flora and fauna detected.

"Two vultures, the white-backed vulture and long-billed vultures are heading for extinction. There has been 90 percent drop in the population of these vultures over just the past three to five years. In the past Keoladeo National Park (area 29 sq.km) supported more than 2000 vultures: it was now having a population of 100 birds. The author mentions about the research investigation of Vibhu Park of Bombay Natural History Museum about the aforesaid and four other protected reserve. Buxa, Jaldapara, Gir, Corbett, Simlipal which had fewer number of vultures. The crisis has galvanised Wildlife Organizations both in India and abroad for investigation and they consider a relative of poultry virus may be cause of vulture death."

[*Source: Sunday Telegraph dt. 26.11.2000 by G.S. Madu and A.S.Gill*]

The genetic resources serve as raw materials for tomorrow's biotechnology. Gene pools in the world constitute the additional genetic resources for the future improvement of crop plants, forest trees as well as new and novel plants of potential use in agriculture, forestry, medicine and industry. UNCED Earth Summit deals with various aspects concerning conservation and sustainable use of all component of biological diversity.

"India is now facing a severe threat for her rich biodiversity and genetic resources from habitat fragmentation. The only way to avert a mass extinction is 'Conservation' – our evolutionary responsibility."

[*Source: M.S. Mondal of B.S.I in Nat. Symposium of Zoo. Soc. 1995*]

Various issues about wanton destruction of Tropical forests was first brought to public notice by the National Academy Sciences (NAS, USA) in their report entitled "Research priorities in Tropical Biology." Later NAS published a book entitled 'Biodiversity' where the editor E.O. Wilson (1988) stressed its importance of exploding human populations responsible for degrading the forest resources: he also found that much biodiversity was being lost through extinction caused due to destruction of forests. According to Raven, 1988, Wildon, 1989. Ehrlich and Wilson, 1991 "Hundreds of books, articles, and reports were written, and scores of symposia, workshops, meetings, and seminars were organized to publicize the fact that the Earth's biological diversity was decreasing. Locally and internationally financed programs to conserve biodiversity were implemented. The promise of these new efforts, however, is tampered by the reality of the situation : Between 1950 and 1990 an estimated 30 to 40 percent of the tropical rains forest disappeared, and with the projected disappearances of an equal amount of forests over the next thirty to fifty years : a quarter of the world's species diversity may vanish forever. The situation in India is serious as may be viewed from the charts and tables drawn in various chapters.

[*Source: Modified from Megadiversity Conservation by Chaudhuri & Sarkar 2002*]

There must not be any misconception about the vulnerable status of the fauna as more than 80 species have been considered threatened to extinction and included in Schedule-I of the Wildlife (Protection) Act of 1972. The Zoological Survey of India considers 25 of these 80 species to be highly endangered, and considers general deadline of the population of each species.

In this connection it is worth mentioning that several species of edible frogs, especially Rana trigina, R. hexadactyla, R. crassa and odd number of frogs were killed in the years 1973-1984. One could imagine the impact of this on the balance of insect pest (prey) in our agricultural field vis-a-vis the likely dearth of food for reptilian species. But this figure of the killing of frogs may be several times higher now. It is a telling example of how various species are facing negative biotic impact.

- India has one-sixth of world human population, highly disproportionate to its land area and available resources.
- Population has had a leap from 238 million in 1901 to 1000 million in 2000 which means a rise of about 300 percent in 100 years.
- India has problems of a serious magnitude: food, shelter, and various other necessities of life.
- Due to rise in population the requirement of various food materials will rise and the consequent demand for birds, reptiles and other animals will increase bringing pressure on reserved forests.
- India has half the world's buffalo population and one-seventh of cattle and goat population which put a tremendous pressure on the natural ecosystem.
- Vast genetic heritage, the animal wealth in India, itself is being lost at an alarming rate.

The breakthrough in biotechnology in general and genetic engineering in particular has created some awareness of livelihood and also some new avenues of employment for the growing population. It is now possible to transfer genes from their wild relations, to domesticated plants and animals. Large scale environmental disturbances have thrown out of gear the evolutionary processes of nature which shapes new species of organisms.

Also steps are being taken to keep extensive protected areas for conservation of flora and fauna especially those are under threat.

The report of the Zoological Survey of India reveals that India is losing at an alarming rate its vast genetic heritage, the animal wealth of the country. Useful efforts have been made to save and protect some endangered species, particularly some mammals like tiger, lion, deer, sambar, elephant and birds. But the concept of conservation should have a holistic approach which involves protection of habitat and biodiversity as every species is interlinked with one another, has coevolved and remains tied together in a web of life. The biological diversity and its components have ecological, genetic, scientific, aesthetic, recreational, cultural, educational, social and economic value which has been depicted on some charts.

The Zoological Survey of India has prepared a list of the estimated number of invertebrate and vertebrate fauna of India broadly indicating their status, although research in many field is still in the initial stage.

The countrywide endangerment of flora and fauna is a phenomenon of the day. It is a product of the continuing use of various natural resources by vast and growing human and cattle population with a growing needs of people for food, fuel and shelter, there is less room for wildlife. Most wildlife species are getting endangered primarily due to loss of habitat and poaching though there are other causes of endangerment.

It is well-known that flightless birds, small reptiles and amphibians are often easily eliminated by people. Trade in animal products such as rhino horns, crocodile skins, in live animals such as parrots, peacock, pigeons, partridges is often responsible for reducing species number. The demand for fish and fish products and the modern harvesting technology has reduced abundance of many species, besides industrial pollution, poisoning of waterways and aquatic life. Dams have cut off migration access to breeding or feeding areas. Alien species introduced to new habitats without controls choke out or eliminate local species by occupying the habitat, competing with them: eating them or by bringing new diseases to which the local species are susceptible.

In the Words of Dr. G. Bertrand, Massachusetts Anderbor Society

"In the end, wild endangered species are no more endangered than we are ourselves by our own population growth, pollution, and the hostilities and conflict brought on by resource competition. Endangered wildlife and our survival are interwined — a fact we are still learning."

India's natural vegetation has undergone drastic changes by various forms of agriculture, forestry and urbanization, although some species are also threatened by natural causes as landslides, floods and droughts. The rate of deforestation has been estimated at 1350 sq.km. per annum out of a total of 6,00,000 sq.km. although only just over half of this can be considered adequately stocked forest lands. Moist forests are degraded rapidly. The widespread effects of this degradation include reduction in rainfall in some areas and flooding in others. At Mahabaleswar region the rainfall decreased from 1000 cm a year to just over 600 cm. The situation is alarming.

Threatened Wildlife

Out of innumerable cases of prevailing stress (threatening on wildlife only a few are mentioned here under due to brevity of space.

S. C. Sharma, Addl. Inspector General of Forests and CHES Management Authority of India writes (1998) about the vulnerable status of Wildlife as,

> "During recent years, there has been a lot of concern about the heavy toll on wild animals being taken by poachers and persons involved in illegal trade. The State Governments are now convinced that their efforts need to be augmented by the Wildlife enthusiasts outside the Government System."

India's Wildlife Population is under Severe Threat of Extinction

Avijit Ghosh's report (refer **The Telegraph**, Calcutta, Sunday, 20th February, 2005) presents a nerve wrecking information which he designates as **"The worst moment of India's Natural History."**

He says during a raid of a Delhi warehouse on January 31, the Delhi police seized a huge magnitude of blood-soaked paws, claws, canines and jaws bones of tigers and leopards packed in cartons. Some big cat skins were stained with blood. "It was like being in slaughter house", says R. Giri the police officer.

Sansar Chand the most wanted wildlife smuggler (Virappan of the North) had in his godown 2 tiger skins, 28 leopard skins, 42 other skins, 14 tiger canines, 3 kg of tiger claws, 10 tiger jaws, 60 kg of tiger and leopard paws and 135 kg of porcupine quills.

Belinda Wright, Executive director Wildlife Protection Society of India (WPSI) says – "He (Sansar Chand) is probably responsible for more tiger and leopard deaths than any other else in the country."

Ghosh reaffirms that organised, illegal trafficking in wildlife products continues to flourish WPSI records that in 1994-2003, 634 tigers, 2336 leopards and 698 others were killed by the poachers.

Environmentalists believe that these seizes represent only a small percentage of the total illegal trade. There is burgeoning markets in China and south east Asia where these animals parts are used for clothes, balms, aphrodisiacs, charms and accessories. "Every body part of a tiger is invaluable. Even other body parts of a tiger is invaluable, even its blood and penis", says wildlife lawyer Sudhir Mishra.

The poachers are active all over the country; death of elephant in Corbett National Park by poisoning indicated involvement of poachers from the North east.

About justice it is said that judicial convictions relating to wildlife cases are abysmally low. Out of 748 cases in India where skins of tigers, leopards or others have been seized, there have been only 14 convictions. The cases also proceed at a snail's pace taking about 8-10 years, before being decided. A designated court for wildlife has more than 250 cases in Delhi alone.

Mishra, the wildlife lawyer cites poor application of law by enforcement agencies lack sensitisation on wildlife issues for the trial court magistrates and the absence of aggressive, enforcement strategies in the courts by forest department officials as reason why wildlife traders and poachers stay in operation and out of jail.

Proposal for setting up a Wildlife Crime Control Bureau (WCCB) is languishing in the Union Ministry of Environment and forest since 1995. "If India needs to save endangered wildlife, the WCCB must become a reality immediately", says Wright.

But then wildlife hardly figures in the official priority list. Wright feels that wildlife authorities do not show adequate interest in solving the issue.

No surprise, poachers stalk the jungle with impurity and illegal wildlife traders continue roam free.

Avijit Ghosh reports:

- **Savage Harvest** – India's largest seizure of illegal wildlife products took place in Khaga in Uttar Pradesh's Fatehpur district in January, 2000. No one has been convicted so far.
- **Broken Promises** – On December 22, 2004, India was suspended from the Convention of International Trade in endangered species reason : India did not introduce CITES legislation on wildlife as promised.
- **Pug Count** – Over 100 years age, about 1,00,000 tigers roamed worldwide. They probably number less than 5,000 now.
- **Money for nothing** – The annual budget for Project Tiger (2003-04) stands Rs. 30.67 crores.
- **Roll Call** – The number of frontline staff totals 4,960. With 1,576 tigers in the 28 reserves, the ratio works out to three staff members looking after each tiger.
- **Conclusion** – Only 14 convictions in 748 wildlife cases.

India's Premier Wildlife Project – "The Project Tiger" is in Jeopardy

The present author who was a project tiger director, then conservator of forests, wildlife and in 1995 a member of a 3 member committee to evaluate the state of Project Tiger reserves in India firmly believes that in the present state of India's socioeconomic and sociopolitical condition and absence of a strong political will wildlife in India, especially the tiger's existence is threatened. Ghosh, quotes Valmik Thapar who believes that not only Siriska tiger, the tigers in Dampa, Namphapa, Indrawati, Srisailam, Palamau and Manas face a grave danger.

The author feels that –

Prior to foresters are blamed for this catastrophe it should be proper to find out if there is lack of political will and lacuna in the provisions of forest act which are at the root of the evil.

"This is the Worst Moment in India's Natural History"

A World Wide Fund for Nature in India (WWF-India) report has confirmed every animal lover's worst fears. The tigers of Sariska are gone forever.

On February 11 and 12, an extensive search by a WWF-India team in the erstwhile tiger strongholds Karnavas, Kalighati, Salopka and Pandupole failed to find any trace of the majestic animal. The report says that both day and night-time searches did not "reveal a single sign or evidence, direct or indirect to indicate the continued presence of tigers in the areas." Last summer, there were 18 tigers in the Rajasthan tiger reserve.

No one is still willing to use the two most dreaded words – dead or killed – directly. But the report says the damage to the tiger reserve is likely to have taken place between July and December, 2004. And, that poaching during this period can be the principal cause for the sudden disappearance of the tiger. "If any tigers remain, their numbers are likely to be small", the report says.

More so, leopards have been seen roaming freely in the areas once the exclusive domain of the tiger in these areas.

The report holds that negligence by the forest staff in regular monitoring is one of the likely reasons for the disaster.

And while it talks about the need to adopt all possible measures to counter poaching, it also deems fit to look inwards for possible involvement of the forest reserve staff. Connivance of lower staff with poachers cannot be ruled out, it says.

The two last reported sightings of the tiger in Sariska were last year: on July 14 at Salopka and on August 26 near Kalighati Chowki.

Tiger expert Valmik Thapar believes that the tiger is in grave danger in several other reserves such as Dampha (Mizoram), Namdhapha (Arunachal Pradesh), Indravati (Chattisgarh), Srisailam (Andhra Pradesh), Palamau (Jharkhand), and Manas (Assam). "It is very difficult to find fresh pugmarks in these reserves", he says. "These are other Sariskas waiting to happen."

Callous officialdom is the prime cause of the mess. According to Thapar, the National Board for Wildlife, the country's top policy-making body for wildlife, hasn't met for the past 17 months. "Finally, it is going to meet on March 17," he says.

And incredible though it may sound, the Rajasthan State Wildlife Board hasn't been convened for six years. Says the tiger man, with a said finaility. "This is the worst moment in India's natural history."

Vivek Menon *et al.*, in his Book *Wildlife Crime* (1998) writes in the Introduction of the book.

"India is an importer, exporter and conduct of Wildlife that enters the $ billion annual global trade. In response to this, the Government of India has set its policy and made its laws quite unambiguously. Most wildlife species in India are protected and it is a crime to kill any of them, with the exception of rats, mice, crows and fruit bats. Many plants are also protected by law. Export and import of all wild animals and their parts and products (except shed peacock feathers) is prohibited and the same is true for more than forty species of plants. Despite all these laws and policies the illegal trade in Wildlife continues to flourish. Just as mere laws do not bring down the incidence of heinous crimes in society, the poaching of animals, importing of plants and their subsequent trade has also to be dealt with firmly. We firmly believe that the time has come to recognise the gravity of situation and try to arrest the cataclysmic decline of species."

Wildlife Crime (Source : Menon *et al.*, 1998)

Several species are mentioned hereunder which are in trade and are now threatened:

Wild ducks, geese, assorted water fowl, jungle fowl, partridges, quails, pheasants, doves, pigeons, rose-ringed parakeet, alexandrin parakeet, blossom-headed parakeet, red breasted parakeet, spotted munia, white rumped munia, horbills, flamingoes, storks, pheasants, cranes, ducks, hawks, falcons, merlin; the source also mentions that between 1970-1976, about 13 million birds were exported from India (80% of which were parakeets and munias). Export was banned in 1990 and domestic trade was banned in 1991 of Rana hexadactyla, R. crassa and R. tigrina the legs of which are protected under Schedule - IV of the Wildlife (Protection) Act. edible nest of Swiftlet (Collocalia, fuciphaga, C. uniculur) and trade in antlers of Chital, Sambar, Swamp deer, Barking deer, Hog deer.

The Himalayan black bear and Sloth bear are killed for their gall bladder. Musk deer is heavily poached in the Himalayas for Kasturi. Elephant is heavily poached for ivory. Rhino horns are poached for horns said to be curing diseases. Tiger and leopard skin, bone, animal fur (The Tibetan antelope or Chiru), Skins of rat-snakes, Cobra, saw scaled viper, Russell's viper, kellbacks, etc. are in trade. All crocodiles, pythons and the yellow monitor lizards are given maximum protection by the CITES. Indian reptiles in trade are – Vipera russelli, Xenochrophis piscator, Ptyas mucosus, Naja naja, Ophiophagus hannah, Python malurus, Varanus bengalensis, Varanus flavescens, Varanus salvator, Star Tortoises (Geochelone elegans), roofed turtle (Kachuga tecta), pond turtle (Melanochelys trijuga), Spiny-tailed lizard (Uromastyre hardwickii), etc.; of the primates rhesus (Macaca mulatta), bonnet macaques (Macaca radiata), Common langur (Presbytes entellus), hoolock gibbon (Hylobates hoolock), Slender loris (Nycticebus coucang and Loris tardigradus), pig-tailed macaque (Macaca nemstrina), etc.

All these are being poached for trade purposes: it is therefore, very convincing that Indian wildlife is facing a crisis for existence.

Threatened Waterbirds of Assam

To highlight the destruction of migratory birds in the country a report is reproduced entitled "Killing of Birds" at Jhatinga in Cachar, Assam.

Hundreds of thousands of rare birds dot the landscape, making for a mesmerising sight. But beneath the facade lie ugly nets cast by hunters.

The scene is the same in marshlands across Upper Assam, which plays host to an astonishing variety of migratory birds every winter.

Wildlife officials admit that hundreds of migratory birds are caught and smuggled out of Upper Assam everyday. However, no one is sure where these avian visitors are taken to and whether they end up as meat on dining tables or trophies on living room walls.

Accompanied by members of a non-government organisation, Jorhat Wildlife warden Santa Sarma recently raided several beels (lagoon) along the Brahmaputra and seized a few hunting nets. However, not a single poacher could be apprehended.

Sarma told The Telegraph that "informers" tipped her about presence of poachers in almost all the beels along the Brahmaputra, but all of them fled before she reached the avian habitats.

"Birds are usually caught after midnight (between 2 am and 3 am), making it difficult for the authorities to catch anyone rehanded", she said.

An organised gang of poachers is belived to be behind the racket, but Sarma does not rule out of the involvement of forest personnel too.

"We did not find a single forest department employee at the Kokliamukh beat office during our visit. There should be at least four personnel present at the beat office at all times", she said.

Divisional forest officer of Jorhat circle, M. N. Durarh, said he had received a complaint, but was yet to conduct a survey.

"I took up this assignment recently and am yet to visit the areas where illegal poaching is rampant", he said.

However, Durarh clarified that the forest department was not taking things lightly. "We are definitely worried and are determined not to spare the people involved in smuggling of rare birds", he said.

The forest official said people living in the vicinity of beels could play a major role in protection of migratory avian species. "We plan to organize public awareness camps soon", he added.

Sarma echoed Duarah's view that villagers should be involved in the campaign to protect birds. But she shed the onus was on the forest department to show the way.

"All beels with a sizeable population of migratory birds should be taken over by the forest department. I have already forwarded my personal suggestions to the forest department", she said.

Eminent zoologist P. C. Bhattacharjee, who has been invited to the 2002 International Onrnithological Congress in Beijing, believes Assam is a "unique bird-watching site" as two "flyways" – the east Assam and the central Assam routes – pass through it.

"No other state in the country has two flyways used by migratory birds", he told the Telegraph.

According to the conservator Anwaruddin Choudhury, 290 species of migratory birds visit Assam every year. Of these, waterfowls comprise the bulk.

Vanishing Vultures Fuel Funeral Rethink

No screeching, no sound of their flapping of wings. An ominous quiet hangs over the Towers of Silence, the Parsi tomb for the dead, at the heart of the financial capital, where fewer and fewer vultures are flying in to feast on the bodies.

With the bird's population taking nosedive in Mumbai and in many other areas in the country, the community is searching for ways to dispose of the rotting bodies without the help of nature's most efficient clean-up bridge.

Behram Contractor, a veteran journalist from the community, said Zoroastrians prohibits burial or cremation of bodies.

"But still, many Parsis may prefer to be buried after death as an alternative; though it may not strictly have the approval of the religion", Contractor, the editor of Tabloid, *Afternoon Despatch*, said.

"It does not really matter what happens to your body once you are gone and burying the body in any case is better than leaving it out to rot", he said.

Instead of burying or cremating their bodies, devout Zorastrians leave their bodies out for vultures at the open-air amphitheatres nestled in 20 hectares of forest on upscale Malabar Hill.

"The community is facing an unprecedented crisis in its 2000 year old history", B. K. Karanjia, a Parsi writer of repute said. There are very few vultures left to do the job, he added.

A 1999, study by the Bombay Natural History Society reported an alarming drop in the vulture population in the country. Some ornithologists attributed this decline to an unidentified virus sweeping across South Asia.

Calcutta-bound Turtles Rescued

Two truckloads of Gangetic turtles, an endangered species under the Wildlife Protection Act, escaped hungry human palates when they were rescued on their way to Calcutta from Kannauj and put back to into Gomti river today.

In the biggest ever catch in the state, forest department officials, along with the railway police, nabbed two trucks with 46 quintals of smuggled turtles, on their way to markets in Bengal and Bangladesh.

However, forest officials are worried that if the smuggling and poaching of the delicious turtles, which is on an unprecedented rise, is not halted immediately, it would spell irreparable doom for the turtles.

Troubled by the "disturbing trend", chief wildlife warden R. L. Singh said: "The demand for such turtles, which is believed to have aphrodisiacal properties, is on an increase and the market has spread from Bengal and Bangladesh to other states also."

What is worrying Singh and Wildlife experts is the fact that killing these turtles would upset the river eco-system as they feed carcass and other pollutants in the river.

The population of this endangered species has nose-dived from 11 lakh in 1990 to mere four lakh this year.

Following this "alarming" development, the International Union for Nature and Natural Resources, a UN body, and the Convention on International Trade of Endangered Species have declared these turtles an endangered species.

Singh said the steep hike in the smuggling of these animals is because of the belief that its meat increases sexual potency. "Though this has not been proved scientifically, we are aware that it sells for as much as Rs. 300 a kg in some places because of this myth", he added.

They have also stationed personnel at railway stations. The department has also started a mission to sensitise school children about the threat posed to these animals and river. "School children have to be taught about them" Singh said, adding. "They also double-up as of formers" [*The Telegraph*: December, 2000].

The diversity in physical environment and climate has made our country an ideal habitat for a rich variety of flora and fauna. India's diverse climatic and edaphic conditions support typical type of numerous plants and animals all over the country. India possesses tropical, sub-tropical, temperate and alpine areas resulting in diversity in flora and fauna. India also possesses Thar desert in Rajasthan. Amongst the numerous animals a check-list of few conspicuous animals is presented below:

Local or Common Name	Zoological Name
MAMMALS	
The Himalayan Black Bear	Selenarctos thibetanus
The Cat Bear or Panda	Ailurus fulgeris
The Common Otter	Lutra lutra
The smooth Indian Otter	L. perspicillata
The clawless Otter	Aonyx cinerea

(Contd.)

Local or Common Name	Zoological Name
The Beach or Stone Marten	Martes folna
The Himalayan Weasel	Mustela sibrica
Ermine	M. erminea
The Black Buck	Antelop cervucabra
Chinkara or Indian Gazelle	Gorella gazella
The Pale Weasel	M. altaica
The Yellow-bellied Weasel	M. kafhian
The stripe-backed Weasel	M. strigidorsa
The Marvelled Pale Cat	Vormela peregusna
The Ratel or Honey Badger	Mellivora capensis
The Long Hedge Hog	Hemiechinus auritus
The Eastern Mole	Talpa micrura
The Indian Short-tailed Mole	T. micrura micrura
White-tailed Mole	T. micrura leucura
The Himalayan Marmot	Mormota bobak
Long-tailed Marmot	M. caudata
The alitopiral/Gerbilles	Tatera indica
The Indian Desert Gerbilles	Meriones nurrlanae
Common Dolphin	Delphinus delphis
Plumbeous Dolphin	Sotalia plumbea
Red-sea Bottle-nosed Dolphin	Tursiops adeuneus
Indian Broad beaked Dolphin	Lagenorhynchus clectra
Little Indian Perpase Dolphin	Neomeris phocaenoide
Short-tailed Mole	Talpa micrura
Indian Porcupine	Hystrix indica
Hodgson's Porcupine	Hystrix hodgsoni
Brush-tailed Porcupine	Altherurus macrourus
Indian Hare	Lepus nigricollis
Cape Hare	L. capensis tibetanus
Arabian Hare	L. arabicus craspidotis
Himalayan Mouse Hare	Ochotone royler
Wild Goat	Carpa hircus
Markhor	C. indica
Thar (Himalayan Thar)	Hemitragus jemlahicus
Thar (Nilgiri)	Hemitragus hylocrius

OTHER MAMMALS

Crestless Himalayan Porcupine	Hystrix hodgsoni
Common Indian Hare	Lepus ruficandalus
Hispid Hare	L. hispidus
Indian Elephant	Elephas maximus

(Contd.)

Local or Common Name	Zoological Name
Great Indian One-horned Rhinoceros	Rhinoceros unicornis
The Gaur	Bos gaurus
The Bengal Barking deer	Muntiacus muntijac vaginalis
The Sambhar	Cervus unicolor
Spotted Deer	Axis axis
Hog Deer	Axis porcinus
Wild Boar	Sus scrofa cristatus
Indian Pangolin	Manis crassicaudata
Chinese Pangolin	M. pentadactyla
Binturong	Arctitis binturong
Small Indian Mungoose	Herpestes auropunctatus
Common Mungoose	H. edwardsii
Crab-eating Mungoose	H. urva
Jackal	Canis aurinus
Indian yellow-throated Marten	Martes flavigula
Brown Ferret Badger H. sp.	Melogale personata
Hog Badger	Arctonyx collaris
Common otter	Lutra lutra monticola
Smoothed-coated Indian Otter	L. perspicillata
Clawless Otter	L. leptonyx
Sloth Bear	Melursus urcinus
Serow	Capricornis sumatrensis
Malayan/Common Tree Shrew	Tapaia glis
The Ruddy Mungoose	Herpetes smithi
The striped-necked Mungoose	H. vitticollis
The Brown Mungoose	H. juscus
The stripped Hyena	Hyaena hyaena
The Wolf	Canis lupus
Goral	Nemor headus goral
Takin	Budorcas taxicolor
Tibetan Antelop	Pantholops hodgsoni
The Dugong	Dugong dugon
The Asiatic Wild Ass	Equns hemionus
Indian Wild Ass	E. hemionus khur
The Banteng/Tsaine	Bos banteng
The Yak	Bos grunniens
Wild Buffalo	Bubulus bubalis
The Shapu/Urial	Ovis oriculals
The Nayan/Great Tibetan Sheep	O. ammon
Marcopolos sheep	O. sp.

(Contd.)

Local or Common Name	Zoological Name
The Bharat/Blue Sheep	Pseudois nabura
The Ibex	Capra ibex
Wild Water Buffalo	Babulus bulalis
Pigmy Hog	Sus salvanius
Indian Fox	Vulpes bengalensis
The Gangetic Dolphin	Platanista gangetica
Andaman Wild Pig	Sus scrofa andamanensis
Swamp Deer	Cervus duvauceli
Thamin	Cervus eldi
Desert Fox	Vulpes bucopus
SQUIRREL	
Kashmir Wooly Flying Squirrel	Eupetaurus cinereus
Orange Himalayan Giant Squirrel	Dremonys lokriah
Himalayan Striped Squirrel	Calloniaurus maccielland
Irrawati Squirrel (Honey-Bellied)	C. pygerythrus lokroides
Golden-backed Squirrel	C. caniceps crumps
Common Giant Flying Squirrel	Petaurista magnificus
Lesser Giant Flying Squirrel	P. elegans
The larger Brown Flying Squirrel	P. philippinensis
Red Flying Squirrel	P. altiwenter
Grey-headed flying Squirrel	P. caniceps
Small Travancore Flying Squirrel	Pelinomys fuscocapillus
Kashmir Flying Squirrel	Hylopetes fimbriatus
Particoloured Flying Squirrel	H. alboniger
Hairy Footed Flying Squirrel	Belomys pearsoni
3-striped Palm Squirrel	Funambulus palmarus
5-striped Palm Squirrel	F. pennanti
Giant Squirrel	Ratufa indica
Grizzled Giant Squirrel	R. macrour
The Malayan Giant Squirrel	R. bicolor
Dusky-striped Squirrel	Funambulus bublineatus
BATS	
Common Flying Fox	Pteropus giganteus
Fulvous Fruit Bat	Rousettus leschensulti leschensulti
Southern Short-nosed Fruit Bat	Cynopterus sphinx sphinx
Allied Horseshoes Bat	Rhinolophus rouxi rouxi
Common Yellow Bat	Scotophilus kukli
Wroughton's Bat	S. temmineki wroughtoni

(Contd.)

Local or Common Name	Zoological Name
Hairy-winged Bat	Marpiocephalus harpia lasyunes
Wall Bat	Myotis mystacinus muricola
Tikell's Bat	Hespirotenus tickelli
Barlequin Bat	Scotophilus ornatus
Painted Bat	Kerivoula picta
Beared Sheath-tailed Bat	Taphazon melanopogor
The Great Himalayan Leaf-nosed Bat	Hiaposideras armiger
Serotine	Pipistrellus coromandra
Common Yellow Bat	Scotophilus heathi
Dobsonstub-nosed Bat	Murina cyclotis
Himalayan Hairy-winged Bat	Harlocephalus harpia lasyurus
Hardaricke Temb Bat	Taphozons longimanus
Himalayan Horse Shoe Bat	Rhinolophus affinis
Eastern Himalayan Little Horseshoes Bat	Rhinolophus connutus
Trifoli Horseshoe Bat	R. trifoliatus
Great Eastern Horseshoes Bat	Myotis siligorensis
Asian Barbastelle	Barbastella leucomelas
Hollons Tube-nosed Bat	Murina buttoni
Indian Vampire Bat	Megaderma lyra
Indian Pigmy Pipistrella	Prepistrellus minus

RATS AND MOUSE

House Mouse	Mus musculus
Edwardis Rat	Rattus edwardsi
Little Himalayan Rat	R. eta
Chestnust Rat	R. fulvescens
White-bellied Rat	R. niviveater
Himalayan Rat	R. nitidus
House Rat	Rattus rattus
Mallards/Large-tothen Rat	Dacuomus mallardi
Bay Bamboo Rat	Cannomys oleracea
Little Indian Field Mouse (Bush Rat)	Colunda ellioti
The Indian Tree Shrew	Anathana elloti
Bhutan Duars Rat	Rattus ratus bhotia
Bengal Mole Rat	Bandicoota bengalensis
Bengal Bandicoot	Bandicoota sp.
The Melad/Soffured Field Rat	Millardia meltada
Indian Field Mouse	Mus booduga

(Contd.)

Local or Common Name	Zoological Name
The Spiny mouse	M. platythrix
The White-tailed wood rat	Rattus blanfordi
Alexandrine Rat	R. alexandrus
The Brown Rat	R. norvegicus
Royle's Vole	Alticola roylei
Sikkim Vole	Pitymys sikkimensis
Murree Vole	Hyperacrius wynnei

PRIMATES

The Bonnet Macaque	Macaca radiata
The Rhesus Macaque	M. mulatta
The Assamese Macaque	M. assamensis
The Stump-tailed Macaque	M. arctoides
The Pig-tailed Macaque	M. nemestrina
The Lion-tailed Macaque	M. silenus
The Capped Langur or Leaf monkey	Presbytis pileatus
The Nilgiri Langur	P. Johni
The Slow Loris	Nycticebus coucang
The Slender Loris	Loris tardigradus
The Golden Langur	Presbytis geei
The Common Langur	Semnopethecus entellus
Barbe's Leaf Monkey	Presbytes barbei
Phayre's Leaf Monkey	Trachipihacus pheyrei
The Crab-eating Macaque	Macaca fascicularis
The Hoolock	Hylobetes hoolock

For further study about many more of animals, lizards, reptiles, amphibians, birds and fishes, trees, shrubs, herbs, climbers, etc. the readers may refer to the books of the authors entitled – 'Biodiversity Endangered' (India's Threatened Wildlife and Medicinal Plants) and 'Megadiversity Conservation' (Flora, Fauna and Medicinal Plants of India's Hot Spots).

Megadiversity and biodiversity are much discussed subjects of present days and to preserve them the foresters' role becomes of prime importance. Large scale afforestation along with scientific management of protected areas and aquatic sites including rivers, nallahs, canals, lakes, ponds and other water-bodies have to be undertaken all over the country and advanced soil conservation works also have to be implemented.

Some Special Notes on Wildlife

Forester's special effort is necessary to protect the animals mentioned below:

NON-HUMAN PRIMATES
World : 58 Genera
Madagascar Island is richest with 23 Genera
India : 15 species (some are endemic over a very small area)
(An overview)
N. E. Assam/Tripura

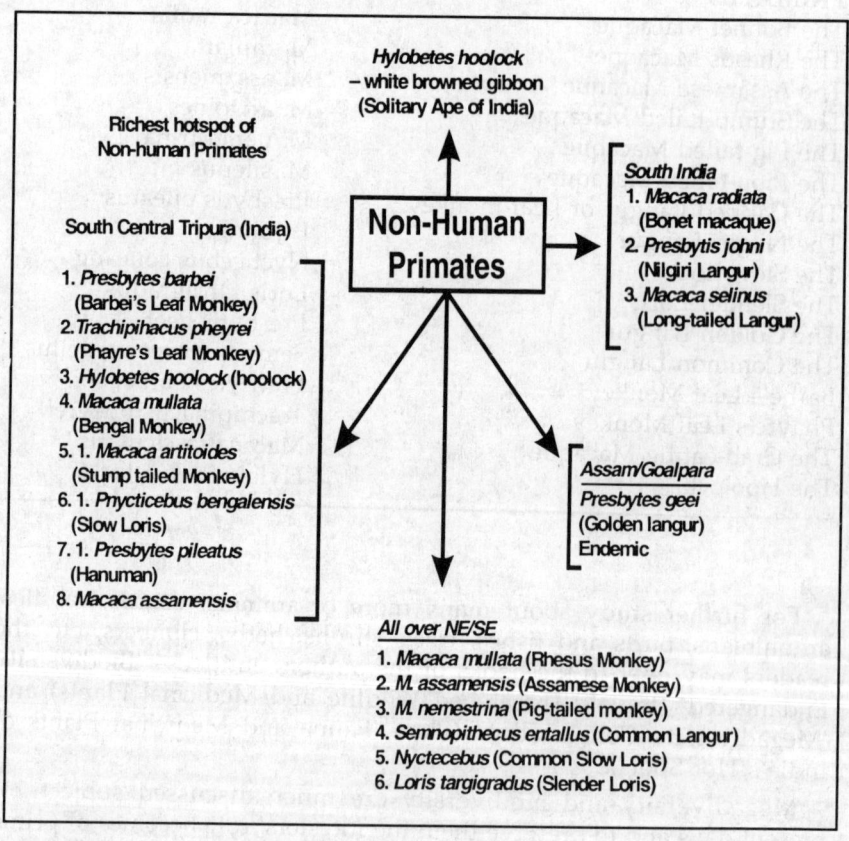

PRIMATES

The order Primates in India has 3 families, 5 genera and 15 species (world has 12 families, 58 genera and 181 species).

Lorisidae – 2 sp. of Loris. Slow Loris and Slender loris.

Cercopithecidae – 12 sp. of Macaca and Presbytis
Hylobatidae – 1 sp. of Hylobatidae.

A Check List has been Prepared which may be Referred to

India does not have Gorilla or Chimpanzee or Orang Utan, but has one ape, a single species, which is Hoolock or white browed Gibbon. This ape exists in Assam, Tripura and other eastern hilly states of Manipur, Mizoram, Nagaland and Arunachal Pradesh. The distribution of apes, monkeys and lemurs are as follows :

- Hoolock – Assam, Tripura, Arunachal Pradesh, Mizoram, Manipur and Nagaland.
- The nonnet macaque – Peninsular India - near Bombay, Godavari river, Trivancore.
- The rhesus macaque - Himalayas, Assam, Central and Northern India, Tripura.
- The Assamese macaque - Himalayan foot hills from Mussoorie eastward, Assam and North east hilly states, Sunderbans, Tripura.
- The Liontailed Macaque - Western Ghat from Canara, Southwards to Kerala, Tamil Nadu.
- The Common Langur - All over India.
- The Capped Langur or Leaf Monkey - Assam and North east hilly states, Tripura.
- The Golden Langur - East of Snakes river at Manas.
- The Nilgiri Langur - Western Ghat from Coong to South-Nilgiri, Anamalai, Brahmagir, Tinnevally, palai hills.
- The Slow Loris - Assam, North east Hills, Tripura.
- The Slender Loris - Southern India, Srilanka.
- Bengal Monkey- Tripura and Eastern India.
- Stumptailed Monkey - Tripura and Eastern states.
- Barbe's Leaf Monkey - Tripura.

Several Species of Primates are in Danger

A small stretch of forest in Tripura from Cherilam Reserve to Trishna Reserve is perhaps the most interesting and vulnerable stretch as the forests are being biotically very much disturbed. The home of golden langur in Manas - Bhutan once is very much disturbed. Three sp., e.g. Bonnet macaque, Liontailed macaque and Nilgiri langur need close protection as these are endemic to the peninsular area. The entire stretch of forests is under heavy depletion.

PANGOLINS

Sangita Mitra of Z.S.I., Calcutta, studied biodiveristy aspects of Indian Pangolin (Manis sp.). She stated that 77 mammals, of 372 mammalian species of India, as vulnerable and endangered as per the Indian Red

Data Book. Pangolin is one such highly specialised species. It's population is not known, but it is rare in the forests. They are killed for the supposed medicinal value of their scales and meat. Though a protected animal included in Schedule-I of Indian Wildlife (Protection) Act (1972) and Appendix-II of CITES' the illegal trade continues. Other biotic and abiotic factors are also causes for the depletion of this species.

STATUS OF REPTILES AND LIZARDS

A BROAD VIEW
BIODIVERSITY DEPLETION

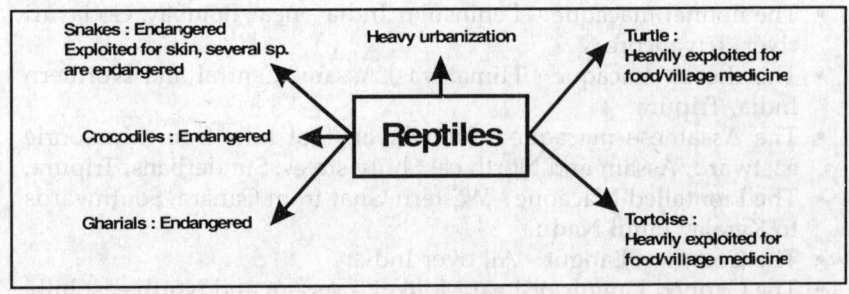

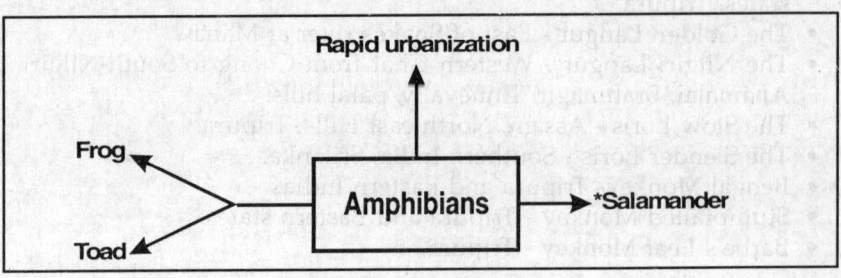

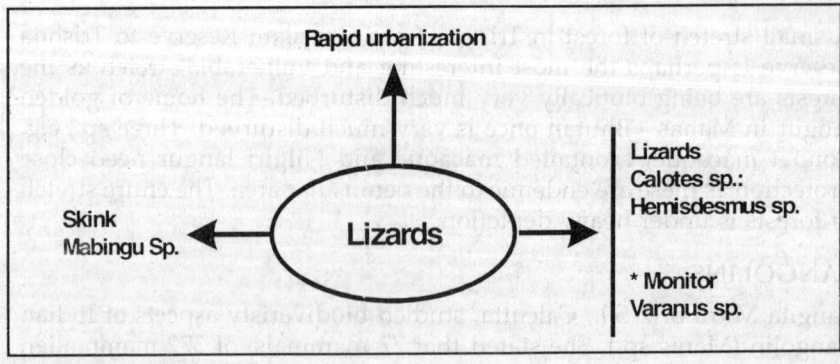

☆ **Endangered**

Perilous Indication of Depletion

Several schedules of Wildlife (Protection) Act, 1972 of India which lists a good number of animals, birds, insects and plants as threatened indicate the sad story of depletion due to various biotic impacts. Some of the species are endangered. More and more species are being enlisted every year that highlights the vulnerability.

REPTILES and LIZARDS (General information)

India is not very rich in these animal species. It has about 52 species of snakes of which 20-30 species may be listed as common, besides another about 30 species of inland turtles and tortoises. About the lizards India has about 158 species. But the status of all these species cannot be easily ascertained as authentic information on individuals of each species absent.

Only evidence (factual) is available in the records of a Government of India Department, which controls the exports and imports of flora and fauna. The records show how innumerable species of flora and fauna and their parts are being exported either legally or illegally. Large number and quantity of such materials protected under Wildlife (Protection) Act, 1972 are being illegally killed and exported which include mammalian skins and bones (Tiger, Leopard, Jackal, Fox, Wolf, hayena, Mongoose, Otter, Porcupine, Deer and several other species). Lizard skins of various species including those of Varanus, Living turtles and tortoises, snakes (by thousands) and prohibited bird species; besides hairs, quills, bones, flowers, fruits, seeds, roots and other floral products.

On December 20, 2000 two truck loads of gangetic turtles (about 46 quintals) were intercepted in Lucknow which were being transported to Calcutta illegally. Forest officials feel that the population of these endangered animal has nose-dived in lakhs this in 2000 [**Source** : *The Telegraph* **dt. 21.12.2000**]. Such instances on illegal collections of snake skins, animal skins and other parts are very common all over the country. Similar instances of large haul are very often reported in various media.

These records are just the tip of monstrous iceberg of illegal exports which are being carried out surreptitiously from this country. Yet, this shows how vulnerable are these floral and faunal species and their parts to the destructive actions of pilferers.

Lizards

Lizards are an inconspicuous, but abundant component of the fauna of India. There are 155 (approx.) species of lizards in India, representing the Geckos, the Garden lizards, the Skinks and the Monitor lizards.

Geckos are nocturnal lizards that live around houses or on trees or on the forest areas. They have amazing running power on smooth surface. These are only reptiles to have developed a true voice.

The Agamids are a major assemblage of medium-sized ground, rock or tree-dwelling lizards. These are diurnal animals.

Skinks are a vast assemblage of mostly small lizards, which are recognized by their long and cylindrical bodies and short limbs. They are found both in the plains and the hills. They are secretive and are found under bark, stones and piles of debris. They are mostly striped, cross-barred or spotted.

Monitors are large sized with mobile head, elongated neck, flattened body with laterally compressed tail and snake-like forked tongue.

Biodiversity in Reptiles and Lizards

The ecological and economic functions of reptiles to the people of India is not as great as the major invertebrate groups - birds, fishes and mammals. They were so long not being ruthlessly exploited for food or skin though turtles and tortoises were always a covetable item of food for man and were heavily exploited.

During the later part of last century snake and lizard skins and Tortoise/turtle shells and flesh had been heavily exploited to meet human need all over the country.

No authentic data on their number is available. But their biological role is acknowledged and it is feared that the present trend of heavy exploitation for snake and lizard skins may be the cause for loss of biodiversity of several species such as varanus sp., Python moluras, Ptyas mucosus, Turtles and Tortoises are also threatened due to heavy exploitation.

Several check lists of snakes, turtles and tortoises, lizards, agamids, chameleon, skins, monitor lizard and lacerids are presented; but these are not to be taken as complete list recorded from India. The list have been prepared to show the readers about the rich assemblages of faunal species. No data on their individual number can be presented.

The status of balance in nature due to actions and interactions of various species of amphibia, lizards, snakes, etc. in the biotically and abiotically affected Indian Habitat is very much disputed. It is felt that expansion of settlement, flood, drought, ruthless killing of such species for fun, food and trade have threatened the very existence of all these groups of animals.

AMPHIBIAN BIODIVERISTY

It is difficult to throw light on the status of various species of amphibia. No numerical count of species have been critically made.

The biological role played by various species cannot be ignored as they have direct bearing on the habitat of agricultural field. Bufo, Ramanella, Limnonectes, Rana, Pailantus, Phacophorous and Ichythyophis are the most common and widely occurring genera. Genera of secondary importance are Amolops, Indiana, Micrixalus and Nictibatracks.

Shrinkage of aquatic habitat, use of biocides, ruthless killing of species of edible purposes are responsible for depletion of resources which play a very active role in keeping balance in nature.

Never to be Seen Again

Trends in the status of threatened species: The Red List Indices show that the status birds and amphibians continues to deteriorate. For birds the RLI demonstrates that their status has deteriorated steadily since 1988, which was the year that birds were first completely assessed. A preliminary assessment of amphibians demonstrates similar rates of decline since 1980. However; amphibian species closest to extinction have shown much steeper rate of decline in status.

Population trends are available for 260 Cycads (Cycadopsida, 288 species in total), and of these 79.6 percent (207 species) are declining, 20.4 percent (53 species) are stable and none are considered to be increasing.

Geography of the Red List: Most threatened species occur in the tropics, especially on mountains and on islands. Most threatened birds, mammals and amphibians are located on the tropical continents : Central and South America; Africa, south of the Sahara; and tropical south and southeast Asia. These realms contain the tropical and subtropical moist broadleaf forests that are believed to harbour the majority of the earth's living terrestrial and fresh water species. Therefore, the patterns evident for mammals, birds and amphibians are likely to be representative of most terrestrial taxonomic groups.

The distribution of threatened marine species is poorly known. Initial findings indicate that threatened marine mammals are concentrated in the northern Pacific Ocean and threatened seabirds, chondrichthyan fishes (sharks, rays and chimeras) and seahorses, etc. in the eastern Indian Ocean and southwest and west-central Pacific.

The uneven distribution of threatened species means that a number of countries have a disproportionate number of species at risk of extinction. Countries with a high number of threatened and threatened endemic species include Australia, Brazil, China, Indonesia and Mexico. Other countries or territories holding particularly large numbers of threatened species include Colombia, India, New Caledonia, Peru, South Africa and Vietnam. Additional countries characterized by particularly high proportionate threat in multiple taxa include Madagascar; São Tomè and Principe, and the Seychelles.

Patterns and distribution of threatened species are relatively congruent between taxonomic groups analysed. Differences are primarily driven by underlying range-size distributions among taxonomic groups (e.g., birds tend to have much larger range sizes than amphibians) and by ecological limitations of specific taxa (e.g., birds are better able to disperse over saltwater than amphibians).

Dwindling Wildlife

The author finds it very relevant to reproduce, with some modifications what D. D. Sarkar and he presented the issue in their book entitled, *Biodiversity Endangered*.

The Colonial Govt. fr med rules, law and acts to conserve, yet ruthless annihilation of wildlife continues. In spite of sincere efforts of the foresters destructive processes go on.

The survival of the balance species and population of wildlife are sociopolitical, sociocultural and socioeconomic issues and a strong political will can only save the remaining animals from destruction.

However, the foresters have played the key role in the protection of bulk wildlife species from outright extinction from India.

SOME SPECIES ARE LISTED IN A FEW CHARTS

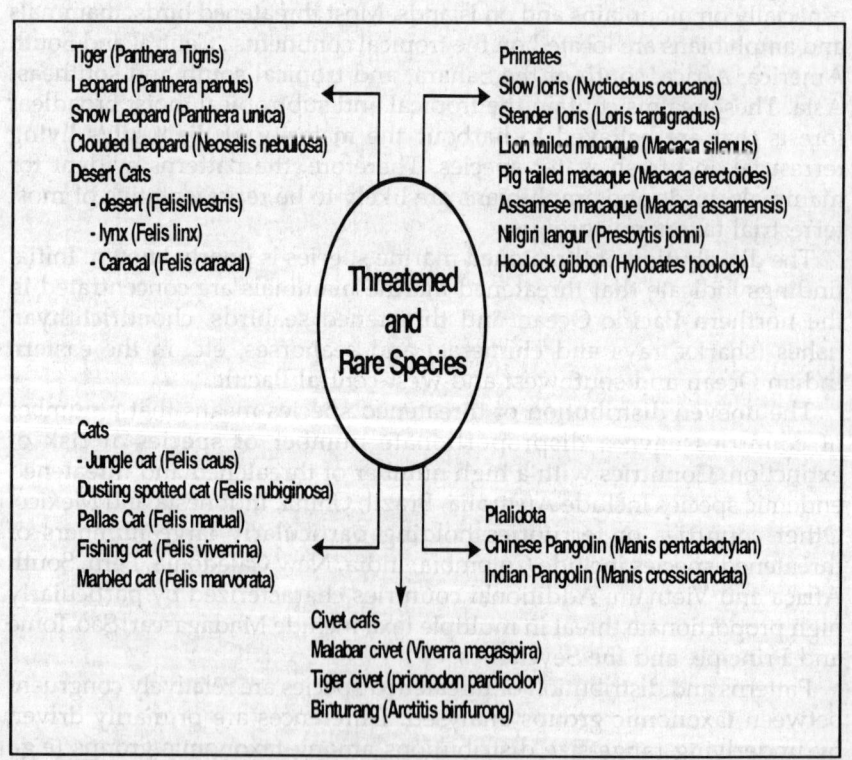

Tiger (Panthera Tigris)
Leopard (Panthera pardus)
Snow Leopard (Panthera unica)
Clouded Leopard (Neoselis nebulosa)
Desert Cats
- desert (Felisilvestris)
- lynx (Felis linx)
- Caracal (Felis caracal)

Primates
Slow loris (Nycticebus coucang)
Stender loris (Loris tardigradus)
Lion tailed mocoque (Macaca silenus)
Pig tailed macaque (Macaca erectoides)
Assamese macaque (Macaca assamensis)
Nilgiri langur (Presbytis johni)
Hoolock gibbon (Hylobates hoolock)

Threatened and Rare Species

Cats
- Jungle cat (Felis caus)
Dusting spotted cat (Felis rubiginosa)
Pallas Cat (Felis manual)
Fishing cat (Felis viverrina)
Marbled cat (Felis marvorata)

Plalidota
Chinese Pangolin (Manis pentadactylan)
Indian Pangolin (Manis crossicandata)

Civet cafs
Malabar civet (Viverra megaspira)
Tiger civet (prionodon pardicolor)
Binturang (Arctitis binfurong)

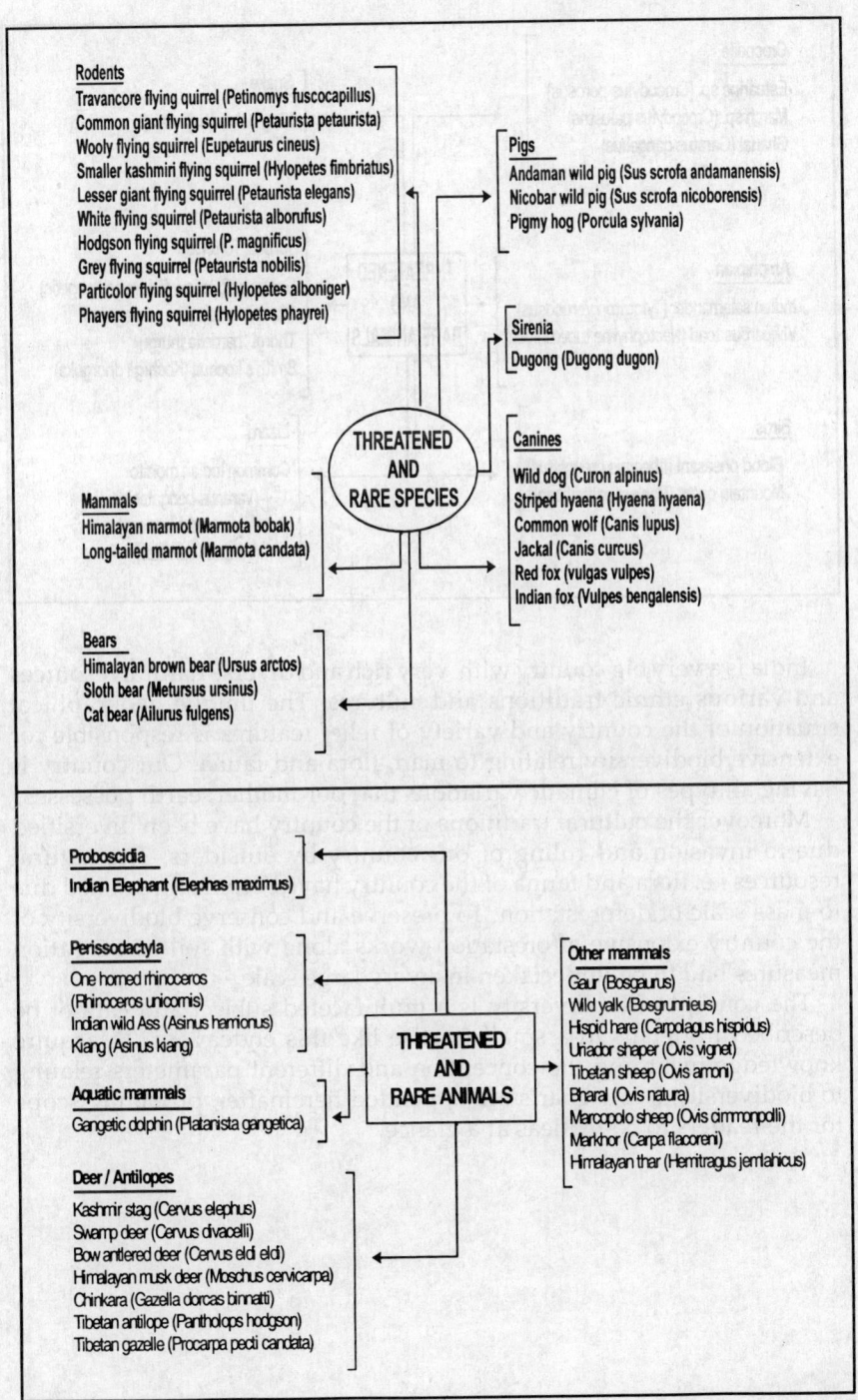

Rodents
Travancore flying quirrel (Petinomys fuscocapillus)
Common giant flying squirrel (Petaurista petaurista)
Wooly flying squirrel (Eupetaurus cineus)
Smaller kashmiri flying squirrel (Hylopetes fimbriatus)
Lesser giant flying squirrel (Petaurista elegans)
White flying squirrel (Petaurista alborufus)
Hodgson flying squirrel (P. magnificus)
Grey flying squirrel (Petaurista nobilis)
Particolor flying squirrel (Hylopetes alboniger)
Phayers flying squirrel (Hylopetes phayrei)

Pigs
Andaman wild pig (Sus scrofa andamanensis)
Nicobar wild pig (Sus scrofa nicoborensis)
Pigmy hog (Porcula sylvania)

Sirenia
Dugong (Dugong dugon)

THREATENED AND RARE SPECIES

Canines
Wild dog (Curon alpinus)
Striped hyaena (Hyaena hyaena)
Common wolf (Canis lupus)
Jackal (Canis curcus)
Red fox (vulgas vulpes)
Indian fox (Vulpes bengalensis)

Mammals
Himalayan marmot (Marmota bobak)
Long-tailed marmot (Marmota candata)

Bears
Himalayan brown bear (Ursus arctos)
Sloth bear (Metursus ursinus)
Cat bear (Ailurus fulgens)

Proboscidia
Indian Elephant (Elephas maximus)

Perissodactyla
One horned rhinoceros
(Rhinoceros unicornis)
Indian wild Ass (Asinus hamionus)
Kiang (Asinus kiang)

THREATENED AND RARE ANIMALS

Other mammals
Gaur (Bosgaurus)
Wild yalk (Bosgrunnieus)
Hispid hare (Carpolagus hispidus)
Uriador shaper (Ovis vignet)
Tibetan sheep (Ovis amoni)
Bharal (Ovis natura)
Marcopolo sheep (Ovis cimmonpolli)
Markhor (Carpa flacoreni)
Himalayan thar (Hemitragus jemlahicus)

Aquatic mammals
Gangetic dolphin (Platanista gangetica)

Deer / Antilopes
Kashmir stag (Cervus elephus)
Swamp deer (Cervus divacelli)
Bow antlered deer (Cervus eldi eldi)
Himalayan musk deer (Moschus cervicarpa)
Chinkara (Gazella dorcas binnatti)
Tibetan antilope (Pantholops hodgson)
Tibetan gazelle (Procarpa pecti candata)

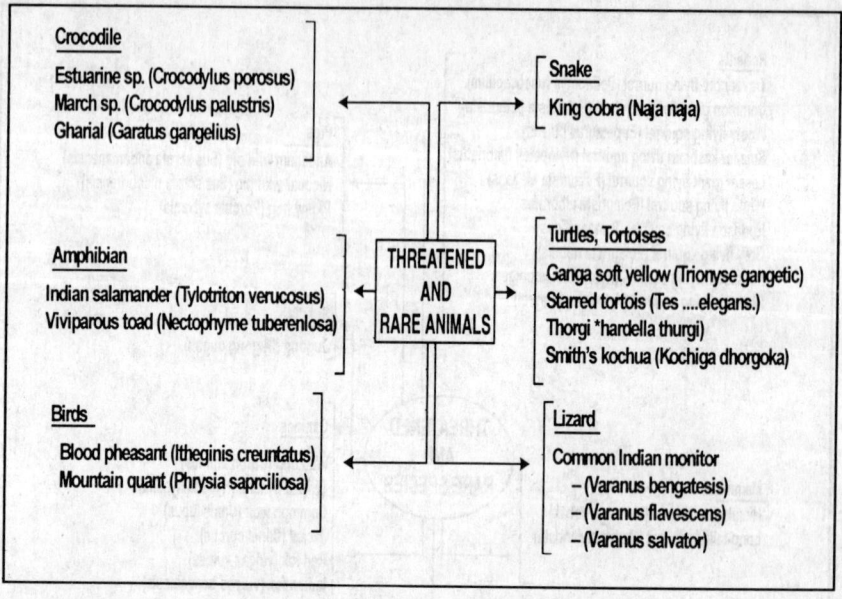

India is a very big country with very rich and diverse natural resources and various ethnic traditions and cultures. The unique geographical situation of the country and variety of relief features is responsible for extensive biodiversity relating to man, flora and fauna. Our country is having all types of climatic variations that our mother earth possesses.

Moreover the cultural traditions of the country have been diversified due to invasion and ruling of our country by outsiders. The natural resources i.e. flora and fauna of the country have become threatened due to mass scale of deforestation. To preserve and conserve biodiversity of the country extensive afforestation works along with soil conservation measures had to be undertaken in a very large scale.

The concept of biodiversity is a multifaceted subject and cannot be described in details in a small treatise like this endeavour. To acquire knowledge regarding the conception and different parameters relating to biodiversity, a few charts are appended hereinafter, providing scope for the readers to have ideas at a glance.

CHART 1: MEGADIVERSITY (FOREST ECOSYSTEM)— AN INDIAN SCENARIO

Forest have diverse life forms and a rich number of genera and species. This diagram shows various life forms. As there is no scope to give details of each form or to present a check list due to space restrictions, each form will be discussed in short outlines. India has 45,000 plant species.

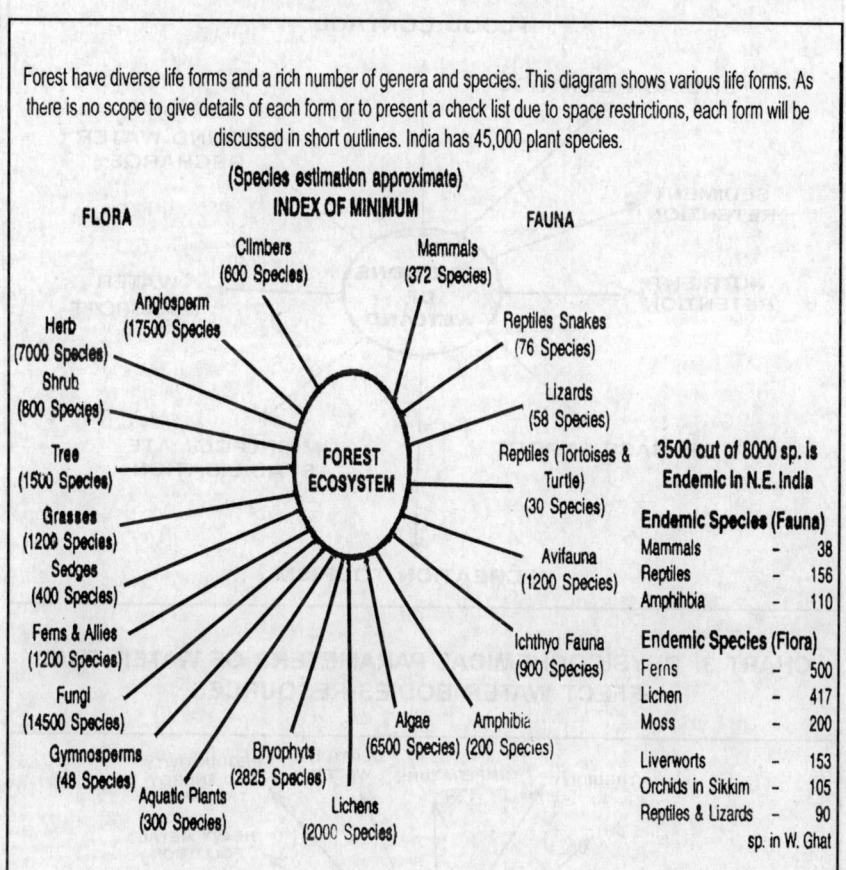

(Species estimation approximate)
INDEX OF MINIMUM

FLORA

FAUNA

Climbers (600 Species)
Mammals (372 Species)

Angiosperm (17500 Species)
Reptiles Snakes (76 Species)

Herb (7000 Species)
Lizards (58 Species)

Shrub (800 Species)

Tree (1500 Species)
Reptiles (Tortoises & Turtle) (30 Species)

Grasses (1200 Species)

Sedges (400 Species)
Avifauna (1200 Species)

Ferns & Allies (1200 Species)

Fungi (14500 Species)
Ichthyo Fauna (900 Species)

Gymnosperms (48 Species)
Algae (6500 Species)
Amphibia (200 Species)

Bryophyts (2825 Species)

Aquatic Plants (300 Species)
Lichens (2000 Species)

FOREST ECOSYSTEM

3500 out of 8000 sp. is Endemic in N.E. India

Endemic Species (Fauna)

Mammals	–	38
Reptiles	–	156
Amphibia	–	110

Endemic Species (Flora)

Ferns	–	500
Lichen	–	417
Moss	–	200
Liverworts	–	153
Orchids in Sikkim	–	105
Reptiles & Lizards	–	90

sp. in W. Ghat

CHART 2: FUNCTIONS OF WATER BODIES

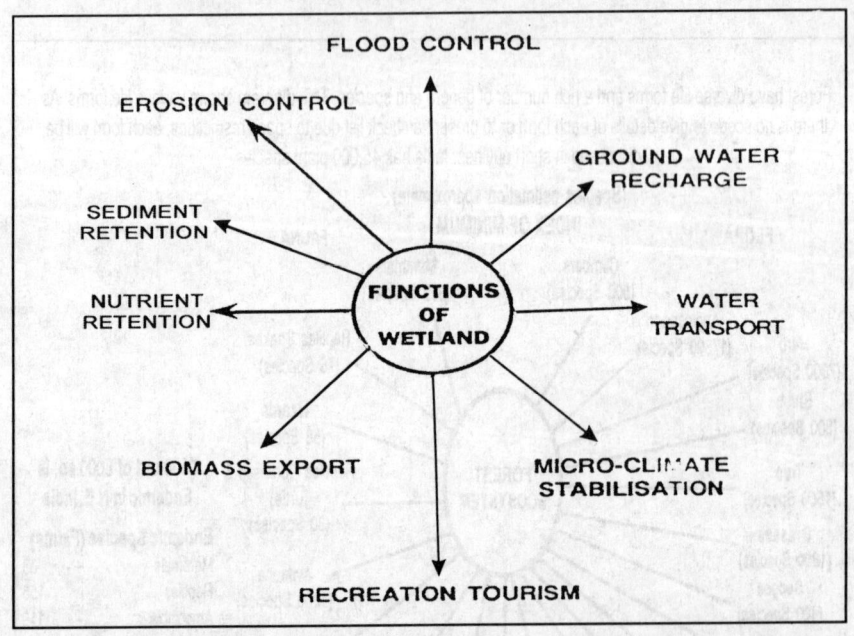

CHART 3: PHYSICOCHEMICAL PARAMETERS OF WATER THAT AFFECT WATER BODIES RESOURCES

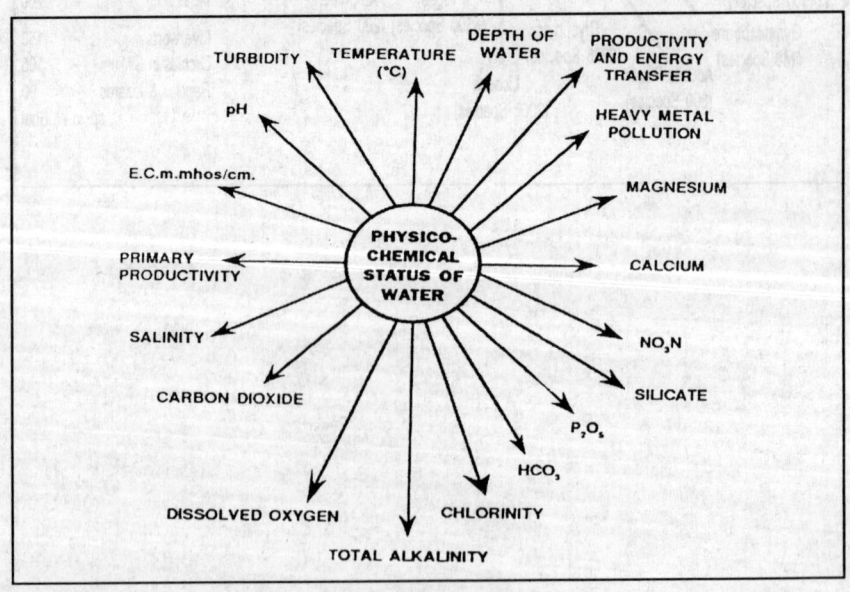

CHART 4: FLORAL ENVIRONMENT (ECOLOGICAL NICHE) AND RESOURCES

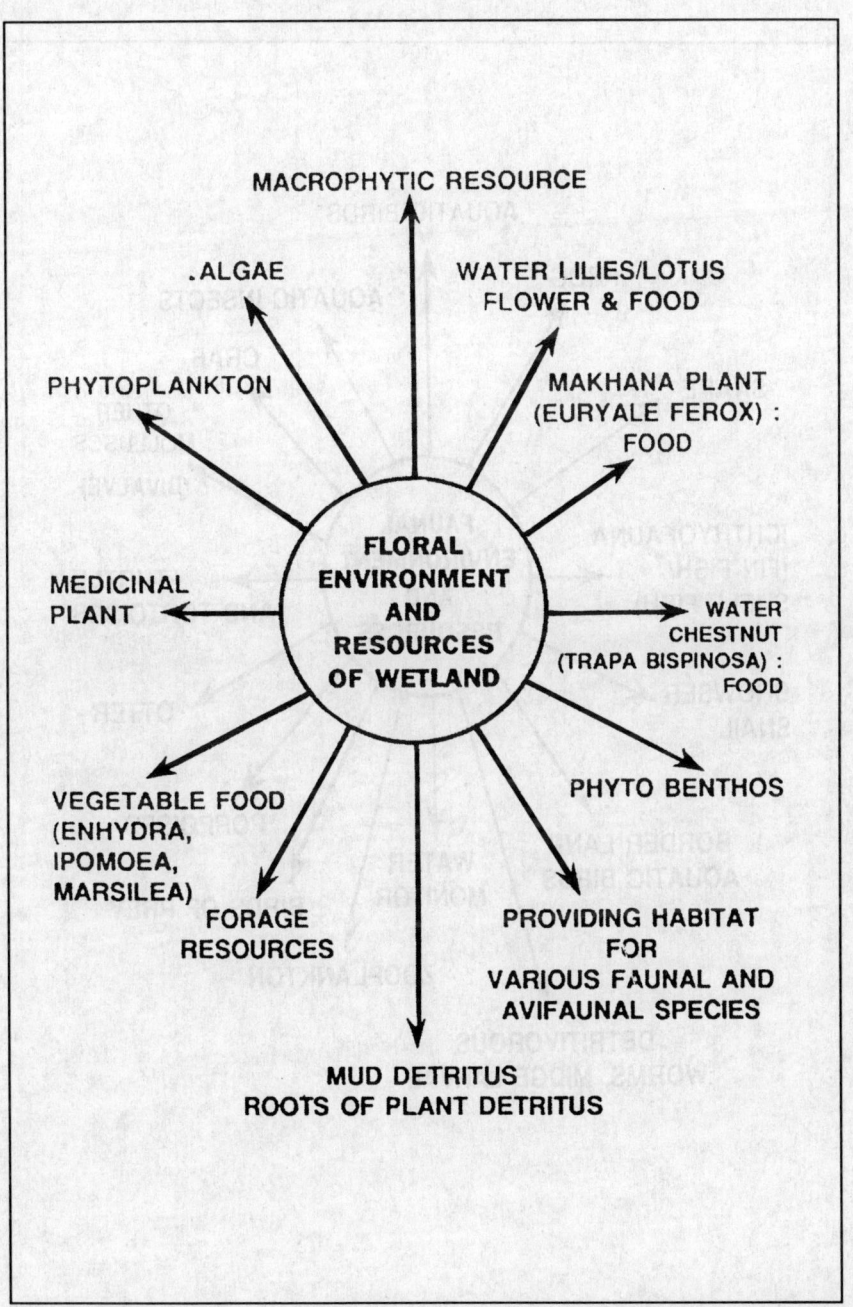

MACROPHYTIC RESOURCE

. ALGAE

WATER LILIES/LOTUS FLOWER & FOOD

PHYTOPLANKTON

MAKHANA PLANT (EURYALE FEROX) : FOOD

MEDICINAL PLANT

FLORAL ENVIRONMENT AND RESOURCES OF WETLAND

WATER CHESTNUT (TRAPA BISPINOSA) : FOOD

VEGETABLE FOOD (ENHYDRA, IPOMOEA, MARSILEA)

PHYTO BENTHOS

FORAGE RESOURCES

PROVIDING HABITAT FOR VARIOUS FAUNAL AND AVIFAUNAL SPECIES

MUD DETRITUS ROOTS OF PLANT DETRITUS

CHART 5: FAUNAL ENVIRONMENT (ECOLOGICAL NICHE) AND RESOURCES

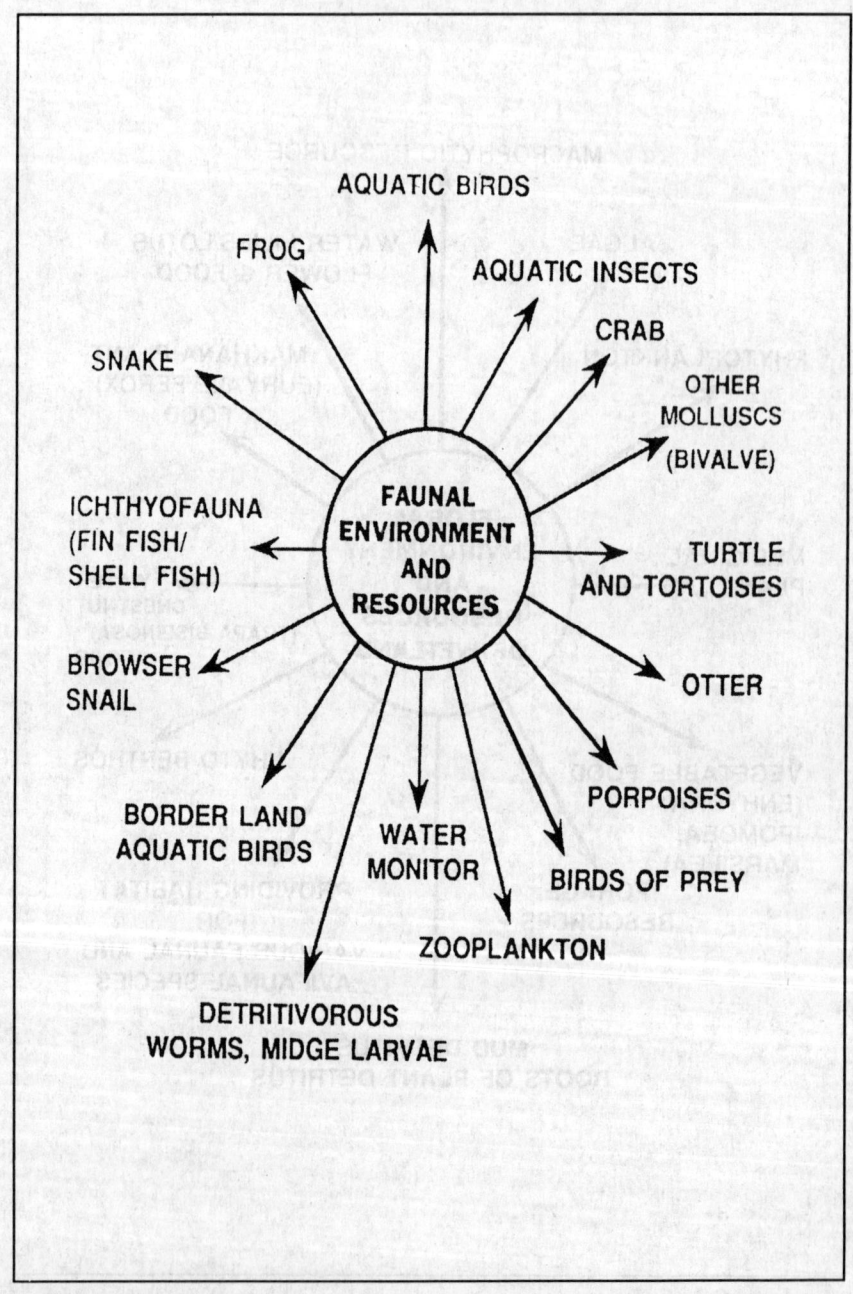

CHART 6: POND WEB

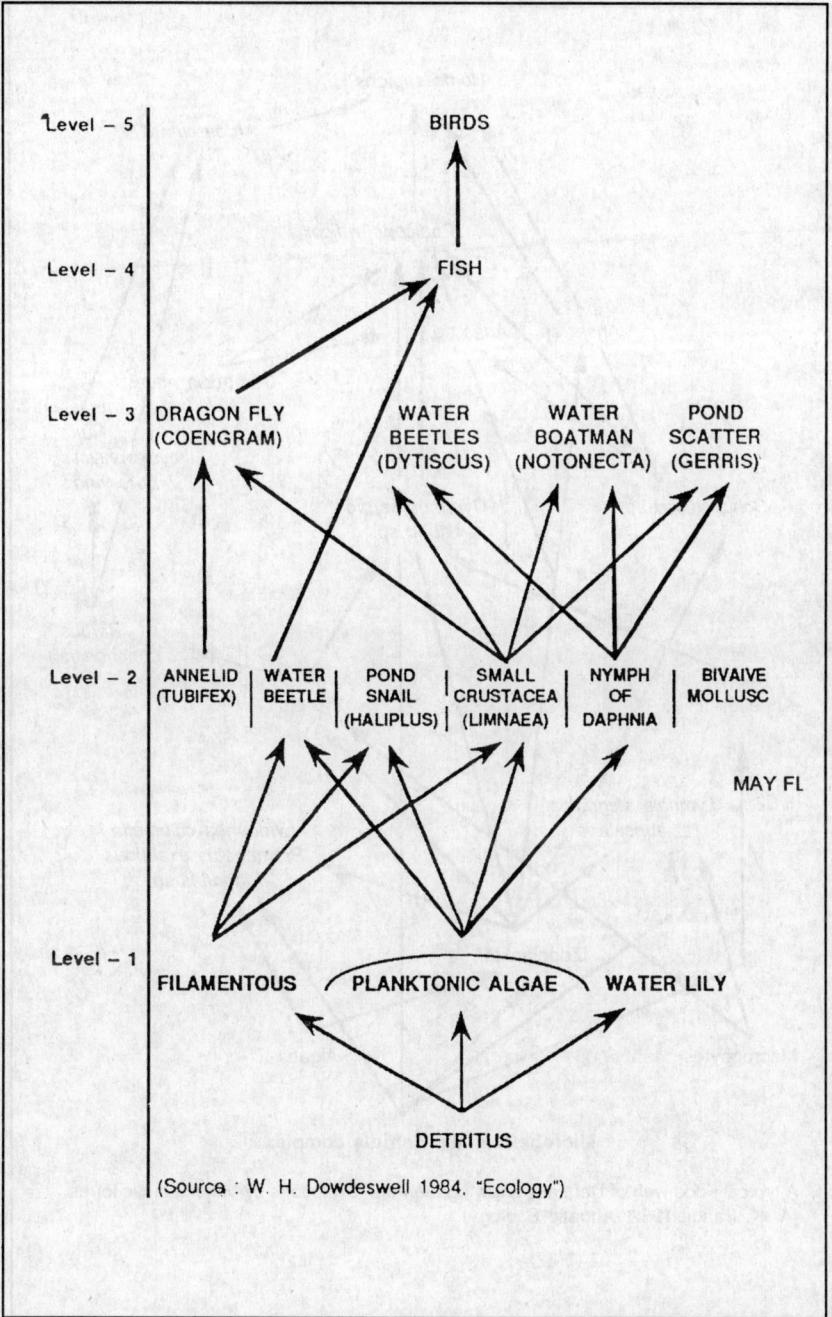

(Source : W. H. Dowdeswell 1984, "Ecology")

CHART 7: FOOD WEB IN A LAKE

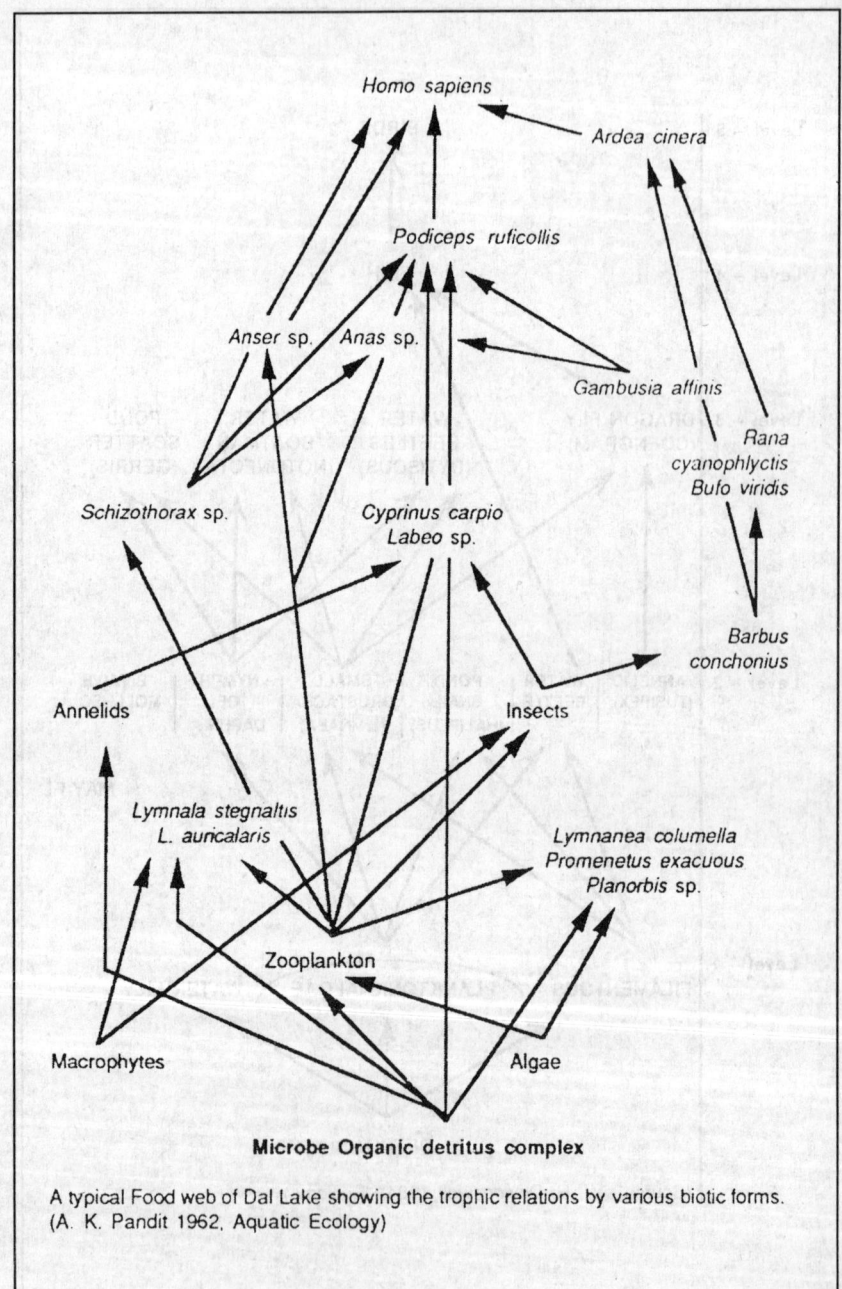

A typical Food web of Dal Lake showing the trophic relations by various biotic forms.
(A. K. Pandit 1962, Aquatic Ecology)

CHART 8: THREATS TO INDIA'S NATURE

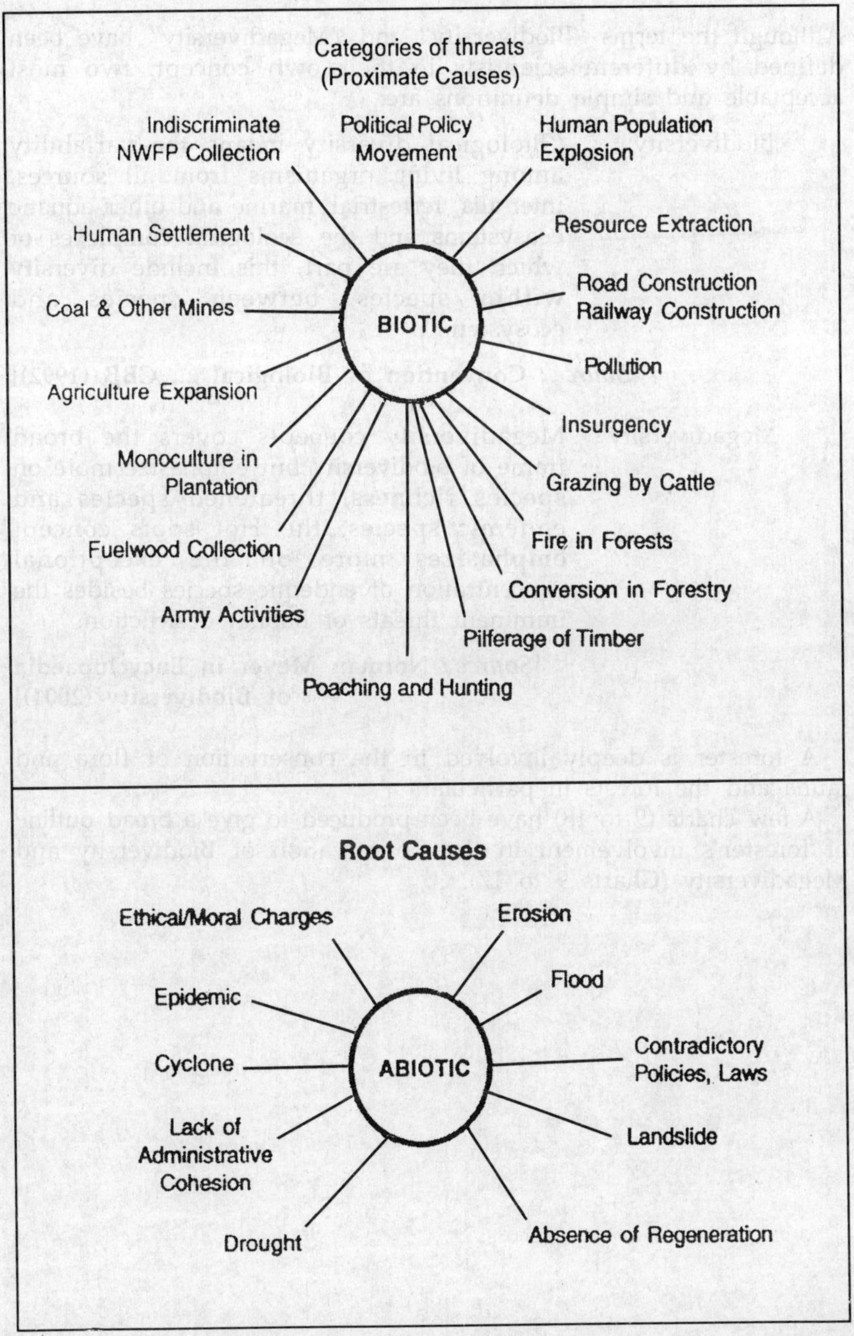

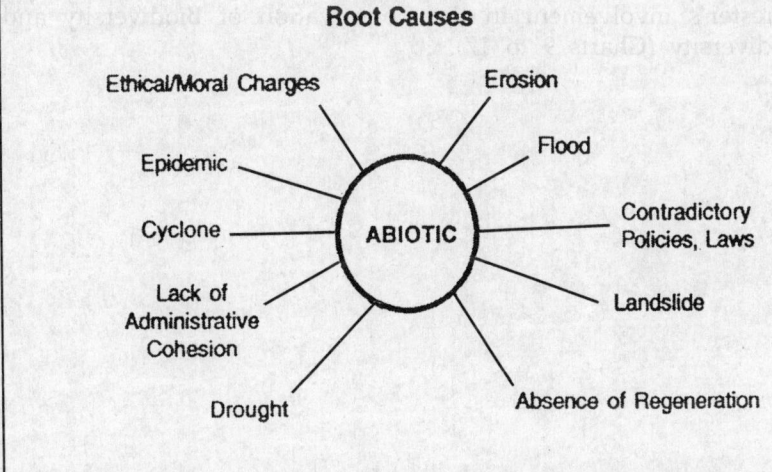

Categories of threats
(Proximate Causes)

Indiscriminate NWFP Collection

Political Policy Movement

Human Population Explosion

Human Settlement

Resource Extraction

Coal & Other Mines

Road Construction Railway Construction

BIOTIC

Pollution

Agriculture Expansion

Insurgency

Monoculture in Plantation

Grazing by Cattle

Fuelwood Collection

Fire in Forests

Army Activities

Conversion in Forestry

Pilferage of Timber

Poaching and Hunting

Root Causes

Ethical/Moral Charges

Erosion

Epidemic

Flood

Cyclone

ABIOTIC

Contradictory Policies, Laws

Lack of Administrative Cohesion

Landslide

Drought

Absence of Regeneration

Forester's Involvement in Biodiversity and Megadiversity Conservation

Although the terms "Biodiversity" and "Megadiversity" have been defined by different scientists in their own concept, two most acceptable and simple definitions are:

Biodiversity : "Biological diversity means the variability among living organisms from all sources, inter-alia, terrestrial, marine and other aquatic ecosystems and the ecological complexes of which they are part, this include diversity within species, between species and ecosystem."

[*Source :* **Convention of Biological .... CBB (1992)**]

Megadiversity : Megadiversity concepts covers the broad frame of biodiversity but emphasize more on species richness, threatened species and endemic species; the Hot spots concept emphasizes more on the exceptional concentration of endemic species besides the imminent threats of habitat destruction.

[*Source :* **Norman Meyer in Encyclopaedia of Biodiversity (2001)**]

A forester is deeply involved in the conservation of flora and fauna and the forests in particular.

A few charts (9 to 18) have been produced to give a broad outline of forester's involvement in the Conservation of Biodiversity and Megadiversity (Charts 9 to 17).

CHART 9: BIODIVERSITY AND MEGADIVERSITY THAT CONCERN A FORESTER

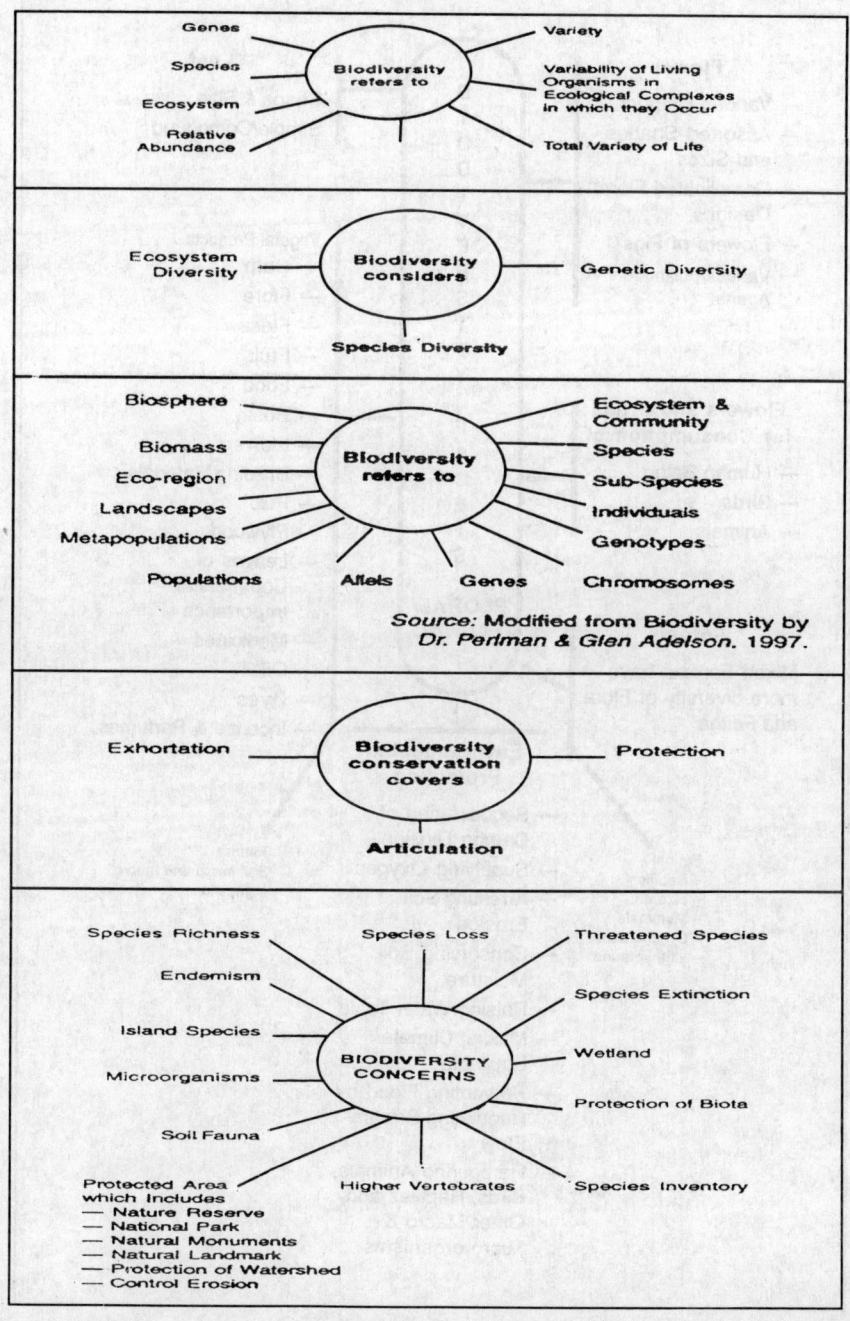

Biodiversity refers to
- Genes
- Species
- Ecosystem
- Relative Abundance
- Variety
- Variability of Living Organisms in Ecological Complexes in which they Occur
- Total Variety of Life

Biodiversity considers
- Ecosystem Diversity
- Genetic Diversity
- Species Diversity

Biodiversity refers to
- Biosphere
- Biomass
- Eco-region
- Landscapes
- Metapopulations
- Populations
- Allels
- Genes
- Ecosystem & Community
- Species
- Sub-Species
- Individuals
- Genotypes
- Chromosomes

Source: Modified from Biodiversity by *Dr. Perlman & Glen Adelson*. 1997.

Biodiversity conservation covers
- Exhortation
- Protection
- Articulation

BIODIVERSITY CONCERNS
- Species Richness
- Endemism
- Island Species
- Microorganisms
- Soil Fauna
- Protected Area which Includes
  — Nature Reserve
  — National Park
  — Natural Monuments
  — Natural Landmark
  — Protection of Watershed
  — Control Erosion
- Species Loss
- Higher Vertebrates
- Threatened Species
- Species Extinction
- Wetland
- Protection of Biota
- Species Inventory

CHART 10: FACETS OF BIODIVERSITY IN FORESTS

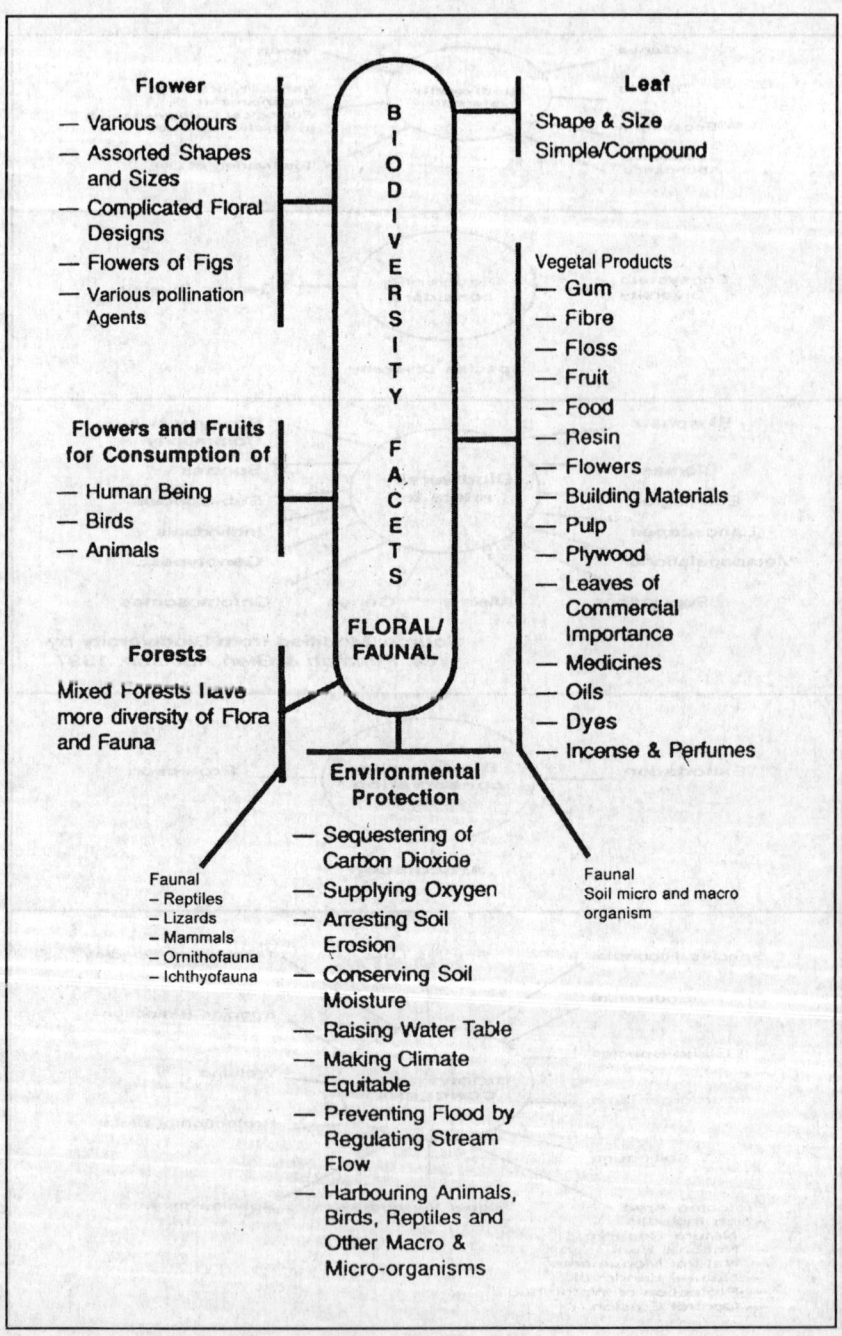

Flower
— Various Colours
— Assorted Shapes and Sizes
— Complicated Floral Designs
— Flowers of Figs
— Various pollination Agents

Flowers and Fruits for Consumption of
— Human Being
— Birds
— Animals

Forests
Mixed Forests have more diversity of Flora and Fauna

BIODIVERSITY FACETS

FLORAL/ FAUNAL

Leaf
Shape & Size
Simple/Compound

Vegetal Products
— Gum
— Fibre
— Floss
— Fruit
— Food
— Resin
— Flowers
— Building Materials
— Pulp
— Plywood
— Leaves of Commercial Importance
— Medicines
— Oils
— Dyes
— Incense & Perfumes

Faunal
– Reptiles
– Lizards
– Mammals
– Ornithofauna
– Ichthyofauna

Faunal
Soil micro and macro organism

Environmental Protection
— Sequestering of Carbon Dioxide
— Supplying Oxygen
— Arresting Soil Erosion
— Conserving Soil Moisture
— Raising Water Table
— Making Climate Equitable
— Preventing Flood by Regulating Stream Flow
— Harbouring Animals, Birds, Reptiles and Other Macro & Micro-organisms

CHART 11: VARIOUS OTHER CONCEPTS OF VALUES OF ENVIRONMENT

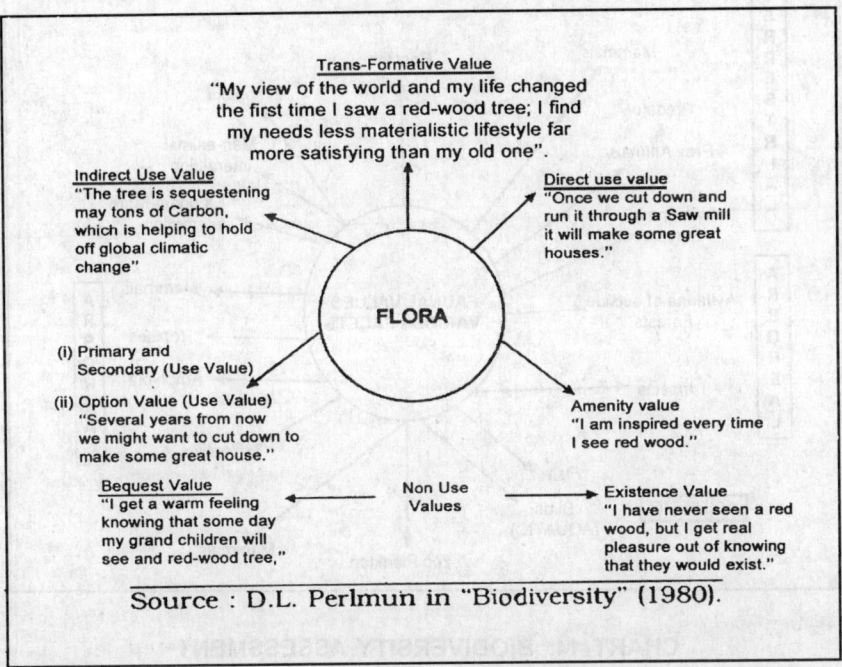

Source : D.L. Perlmun in "Biodiversity" (1980).

CHART 12: VALUING BIODIVERSITY VALUING THE FORESTS IN SUSTAINABLE DEVELOPMENT

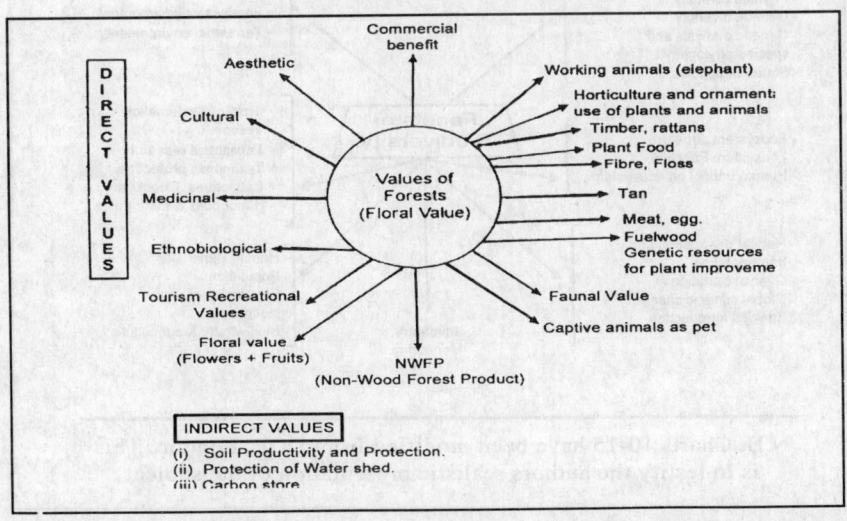

CHART 13: BIODIVERSITY/MEGADIVERSITY VALUES

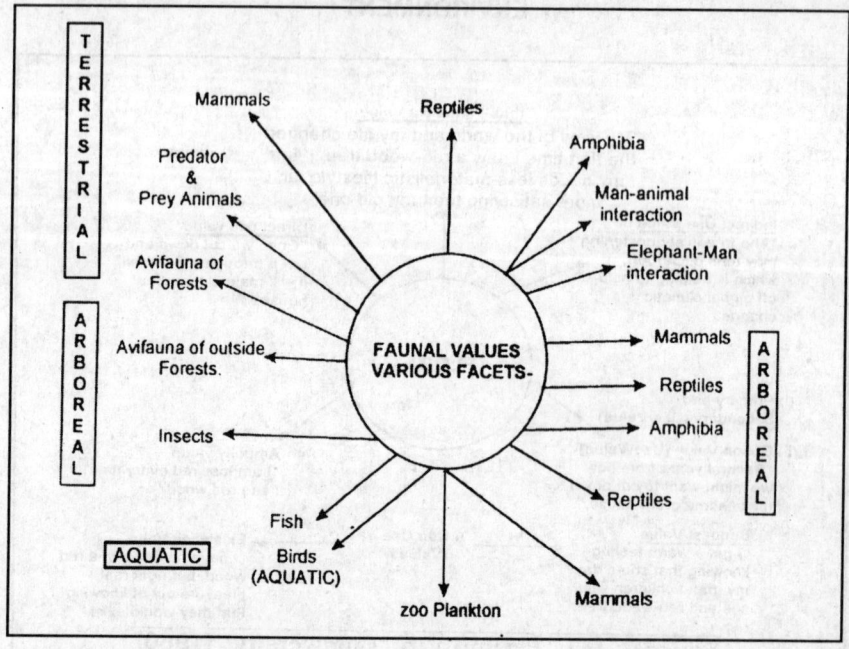

CHART 14: BIODIVERSITY ASSESSMENT

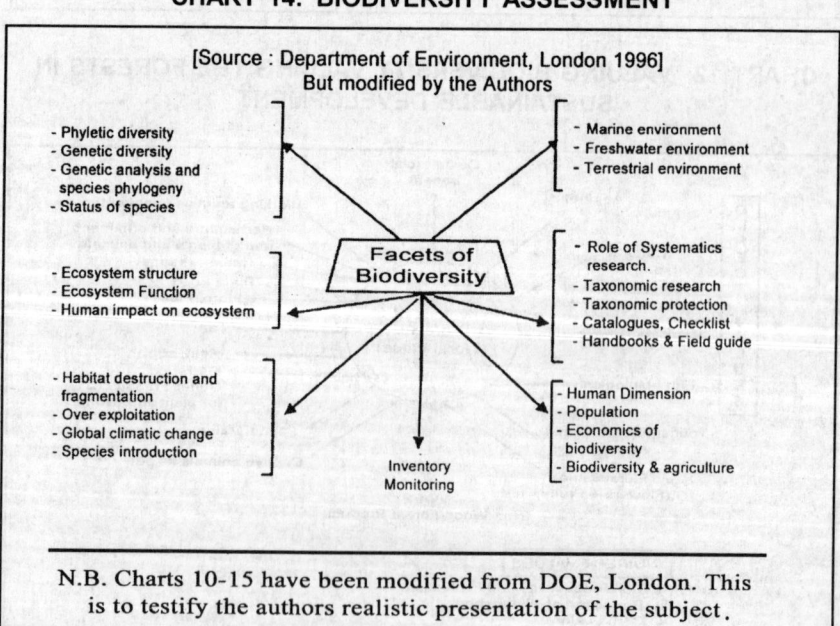

N.B. Charts 10-15 have been modified from DOE, London. This is to testify the authors realistic presentation of the subject.

CHART 15: SHOWING VARIOUS FACETS OF BIODIVERSITY/ MEGADIVERSITY

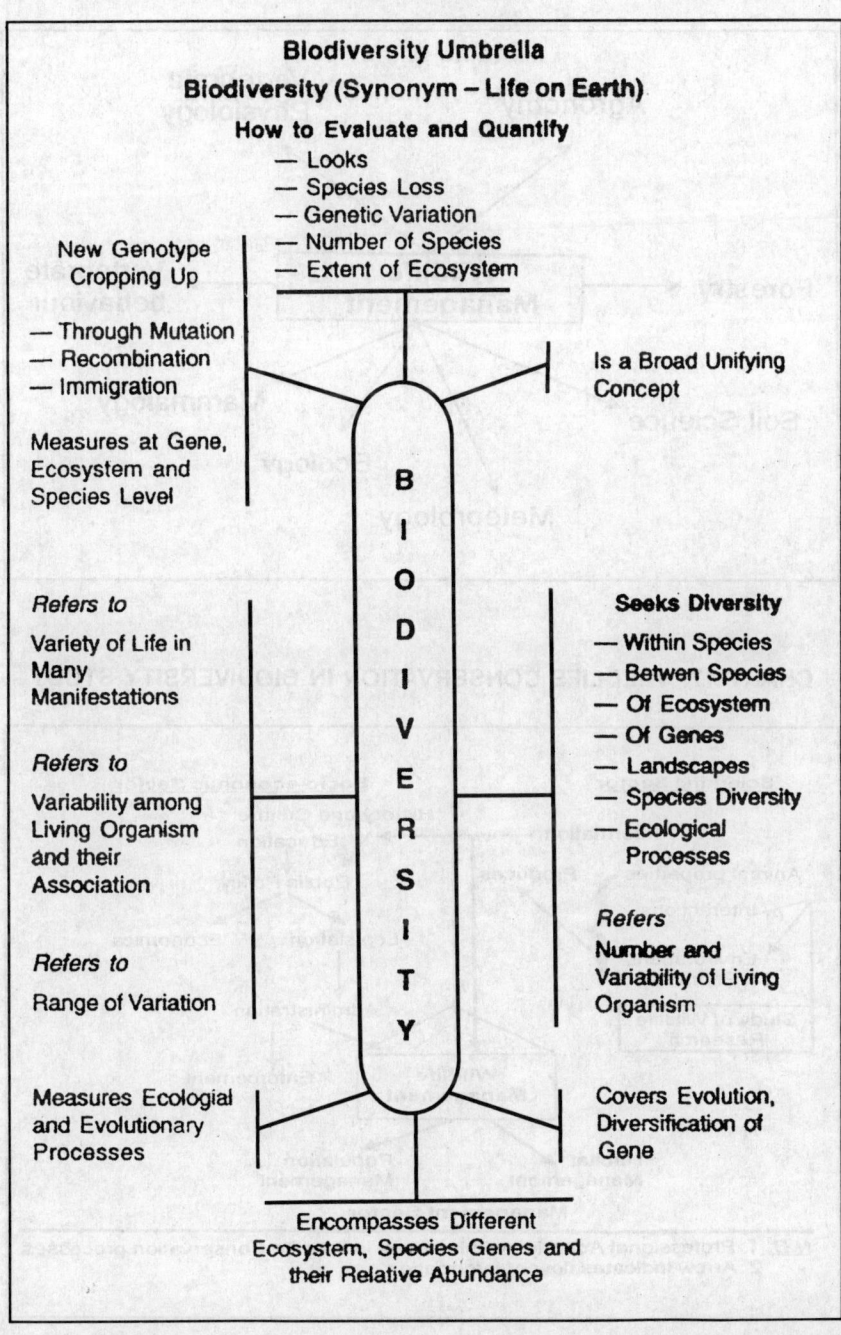

Biodiversity Umbrella

Biodiversity (Synonym – Life on Earth)

How to Evaluate and Quantify
— Looks
— Species Loss
— Genetic Variation
— Number of Species
— Extent of Ecosystem

New Genotype
Cropping Up

— Through Mutation
— Recombination
— Immigration

Measures at Gene,
Ecosystem and
Species Level

Is a Broad Unifying
Concept

Refers to

Variety of Life in
Many
Manifestations

Refers to

Variability among
Living Organism
and their
Association

Refers to

Range of Variation

Seeks Diversity

— Within Species
— Betwen Species
— Of Ecosystem
— Of Genes
— Landscapes
— Species Diversity
— Ecological
    Processes

Refers
Number and
Variability of Living
Organism

Measures Ecologial
and Evolutionary
Processes

Covers Evolution,
Diversification of
Gene

Encompasses Different
Ecosystem, Species Genes and
their Relative Abundance

CHART 16: WILDLIFE MANAGEMENT: ABUNDANCE AND DIVERSITY INFORMATION

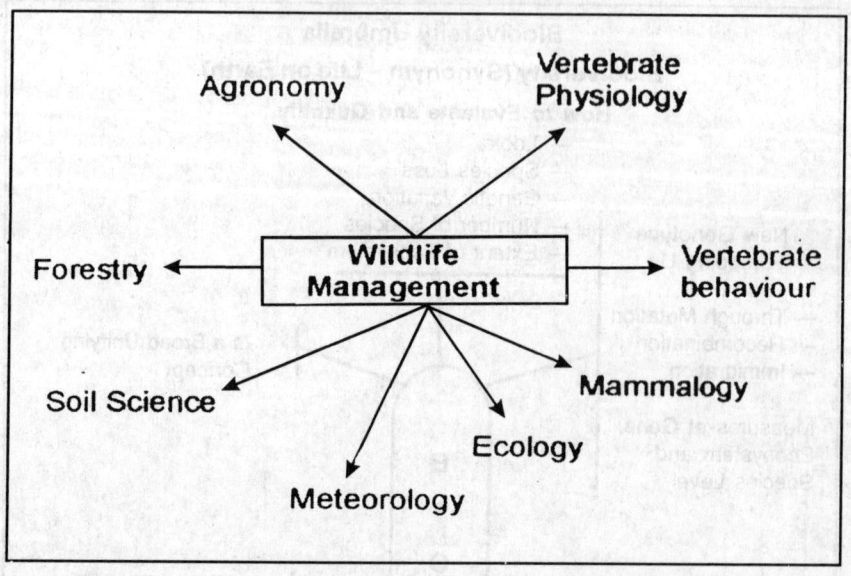

CHART 17: WILDLIFE CONSERVATION IN BIODIVERSITY STUDY

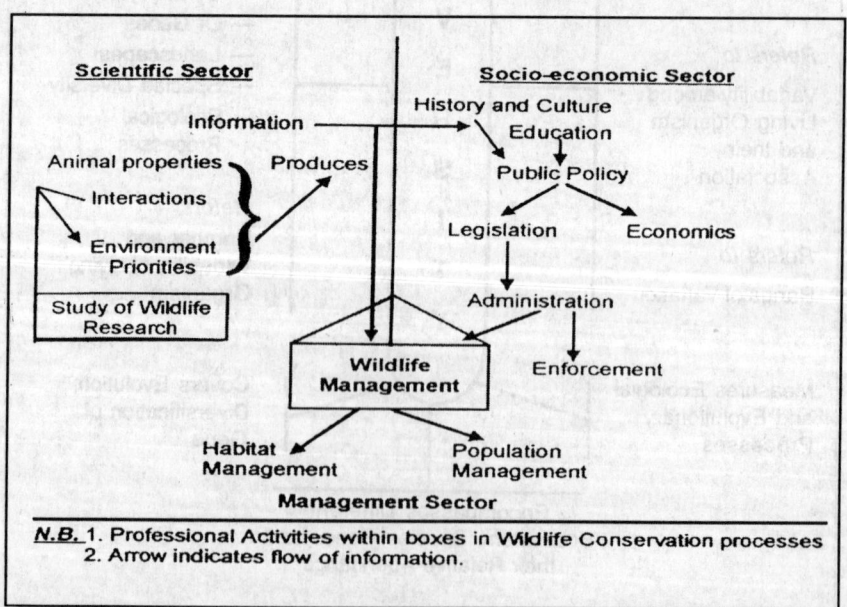

CHART 18: CONSERVATION BIOLOGY IS A MULTIDISCIPLINARY SUBJECT IN DIVERSITY STUDY

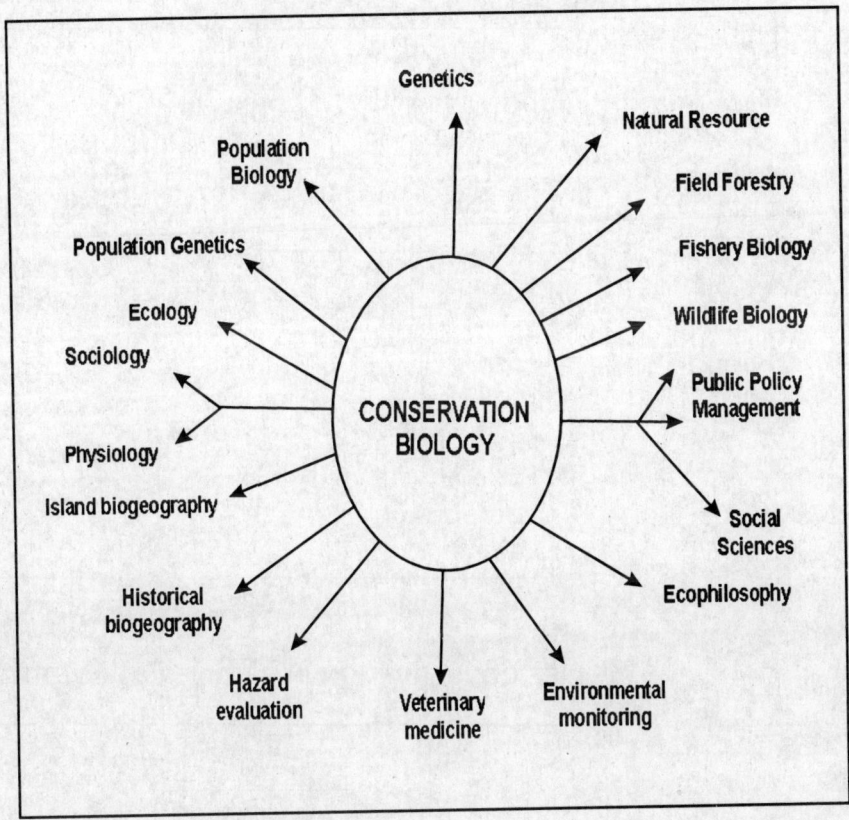

CHART 18: CONSERVATION BIOLOGY IS A MULTIDISCIPLINARY SUBJECT IN DIVERSITY STUDY

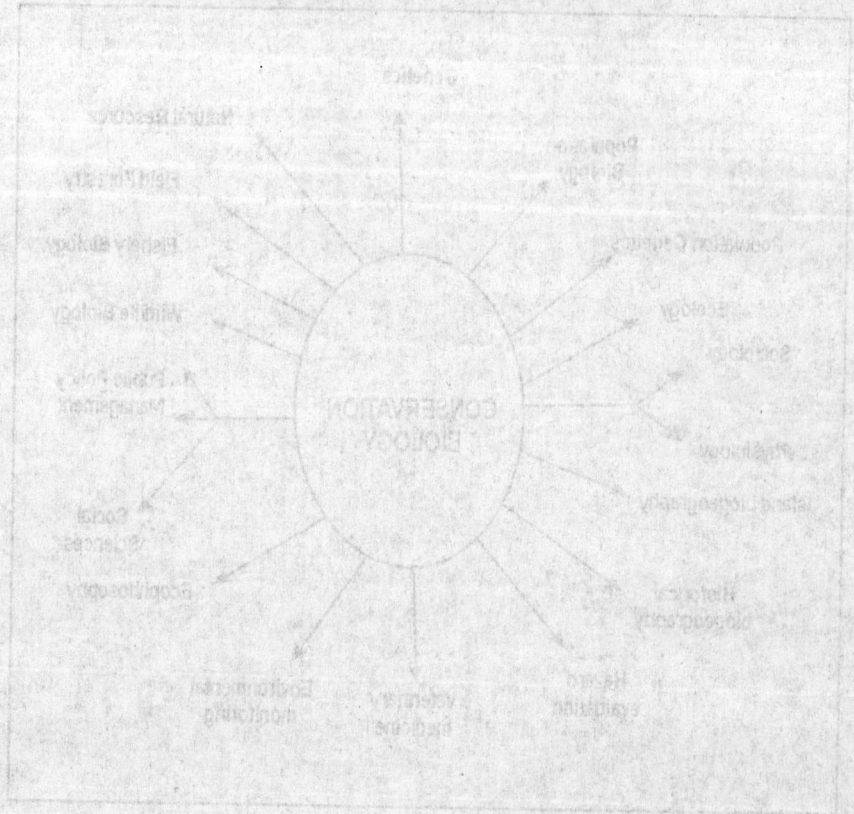

6

India is Facing a Perilous Landuse and Environmental Crisis

— Foresters to Play a Vital Role in Amelioration

The planning of India was done without considering the terrible landuse situation on which the country's future stands. It is like building castles over a soft, cracked and dilapidated foundation that might breakaway anytime like quick sand. The authors drew parallel between the Roman emperor 'Nero' (37-60 A.D) (dream of building new city of Rome while he was playing lyre in full view of burning city) and the planning for development of modern India, the latter having been fully oblivious of landuse situation.

Besides these natural calamities India has to tackle the problems of a huge poverty-stricken, illiterate, unemployed mass where even amenities like safe drinking water, education and medical help, etc. are lacking.

The authors while analysing various facts and figures, documented by the Planning Commission of India, the National Commission of Agriculture and the National Commission on Flood found that they have registered 70% of the total 3.2 m sq. km. of land has been reeling under flood, drought and erosion, etc. Such landuse condition has to be tackled effectively before India takes up a massive, gigantic and long-duration projects. They have stressed that India needs to create an effective cover of plants by massive afforestation mingled with soil conservation engineering works, reclamation of waterbodies– big or small. Here comes the role of foresters who are well-equipped to create a better India. They have training, expertise and the character, built over an experience of about 200 years to rehabilitate depleted soil.

This chapter presents hordes of data on flood, drought, soil, erosion, forest cover, etc. and relevant issues.

6

India is Facing a Perilous Landuse and Environmental Crisis
— Foresters to Play a Vital Role in Amelioration

Date with Doomsday (an Observation by Steve Sawyer, Climate Policy Adviser, Green Peace)

India with a population of just one billion people, is one of the areas most threatened by climate change.

"It's not a pretty picture," said Steve Sawyer, climate policy adviser with Greenpeace in Amsterdam. Global warming and changes to weather patterns are already occurring and there is enough excess carbon dioxide and other greenhouse gases in the atmosphere to drive climatic change for decades to come.

The weather predictions for Asia in 2050 read like a script from a doomsday movie. Except many climatologists and green groups fear they will come true unless there is a concerted global effort to rein in greenhouse gas emissions.

Already, changes are being felt in Asia but worse is likely to come, Sawyer and top climate bodies say, and could lead to mass migration and widespread humanitarian crises.

According to predictions, glaciers will melt faster, some Pacific and India Ocean islands will have to evacuate or build sea defences; storms will become more intense and insect and water-borne diseases will move into new areas as the world warms. All these comes on top of rising populations and spiralling demand for food, water and other resources. Experts say environmental degradation such as deforestation and pollution will likely magnify the impacts of climate change. "The threat to the agricultural base for the Indian subcontinent from drought and increased heat waves, the consequences for the burgeoning Indian economy and the very large number of people to feed are potentially very, very substantial," Sawyer said. Rising sea levels will also bring misery to millions in Asia, he said, causing sea water to inundate fertile rice growing areas and fresh water aquifers, making some areas unhabitable. Sawyer said India and Bangladesh would have to draw up permanent relocation plans for millions of people. "I'm afraid that's almost inevitable." Anwar Ali, a leading climatologist in Bangladesh, says about 15 per cent of the country would be under water if sea levels rose by a metre in the next century.

Perhaps the biggest threat to Asia in the future will be the shortage of clean water. The UN's World Food Programme says Asia accounts for 60 per cent of the world's population but has only 36 per cent of the globe's fresh water. According to the Intergovernmental Panel on Climate Change (IPCC), rapid melting of glaciers poses a major threat to the Indian subcontinent, Southeast Asia and parts of China. Seven major rivers, including the Ganga, Indus, Brahmaputra and the Mekong, begin in the Himalayas and the glacial melt water during summer months is crucial to the livelihoods of hundreds of millions of people downstream. But many of these glaciers are melting quickly and will be unable to act reservoirs that moderate river flows. This means less water in the dry season and the chance of more extreme floods during the wet season.

Sawyer thinks rich countries, by far the biggest polluters, should look after the millions at risk from climate change or suffer the consequences that could include mass migration or trying to feed millions made homeless by droughts and floods in a world struggling to grow enough food. Fears of mass migration have already prompted the Pentagon and the Canadian Security Intelligence Service, among others, to study the risk from climate-induced mass migration. The Pentagon in its 2003 report looked at what might happen if the climate changed abruptly. The result near anarchy. "As global and local carrying capacities are reduced, tensions could mount around the world", it said. This could lead to some wealthier nations becoming virtual fortresses to preserve their resources. "Less fortunate nations, especially those with ancient enmities with their neighbours, many initiate struggles for access to food, clean water, or energy", the report said.

Few places are more exposed to climate change than the low-lying Maldives islands, to the west of Sri Lanka, where the highest natural point is under 2.5 metres. "We still face the threat of sea level rises," Maldivian President Mamoon Abdul Gayoom said in a recent interview. "There is encroachment of the sea on many islands, there is erosion of our beaches", he said. In response, the Maldives is building an island that is a metre higher than the capital Male.

As countries try to adapt, it will be the poor, who suffer most from climate change, said IPCC chairman R. K. Pachauri in a report, *Up in Smoke?*, released last month.

"The impacts of changing climate will fall disproportionately upon developing countries and the poor persons within all countries", he said, meaning the lot of millions of peasants could become far worse than it is now.

[*Source: The Telegraph,* 29/11/2004]

LAND AND AIR QUALITY DEPLETION

A few maps are produced (a very broad-based) on the mining and industrial areas of India which may be superimposed on the drought prone and flood affected areas of India which will testify to the precarious land use structure in India. The maps are appended at the end of this chapter.

Map 1   –   Tribal Population of India, 1991.
Map 2   –   Area under Forest, 1993.
Map 3   –   India : Forest Map of India.
Map 4   –   Map of India showing mineral rich areas.
Map 5   –   Map of India showing selenium polluted areas.
Map 6   –   Map of India showing coal belts.
Map 7   –   Map of India showing energy sites.
Map 8   –   Map of India showing mining areas.
Map 9   –   Map of India showing industrial and mining areas.
Map 10   –   Map of India showing drought prone areas.

A note published on the 29th November, 2004 issue of The Telegraph daily (communicated by Reuters) says

2003   –   A winter cold snap and a summer heat wave killed more than 2000 people.

2004   –   Two thirds of Bangladesh, parts of Nepal and large areas of northeastern India were flooded, affecting 50 million (5 crores) people destroying livelihood and making tens of thousands ill.

DEVASTATION OF LAND AND WATER PROBLEMS AS VIEWED BY B. B. VOHRA

Most useful document dealing with land and water management problem in India is perhaps the one prepared by B. B. Vohra, Addl. Secretary, Ministry of Agriculture published in 1975. The present author thinks that various data presented by him and his opinion about the future remedial measures to be undertaken show Vohra's sincere anxiety at the horrendous land use situation India faces today and the remedy suggested thereof is very relevant to the present issue. As such Vohra's figures have been quoted at a number of places (Vohra was also the Chairman of National Committee on Environmental Planning) as follows:

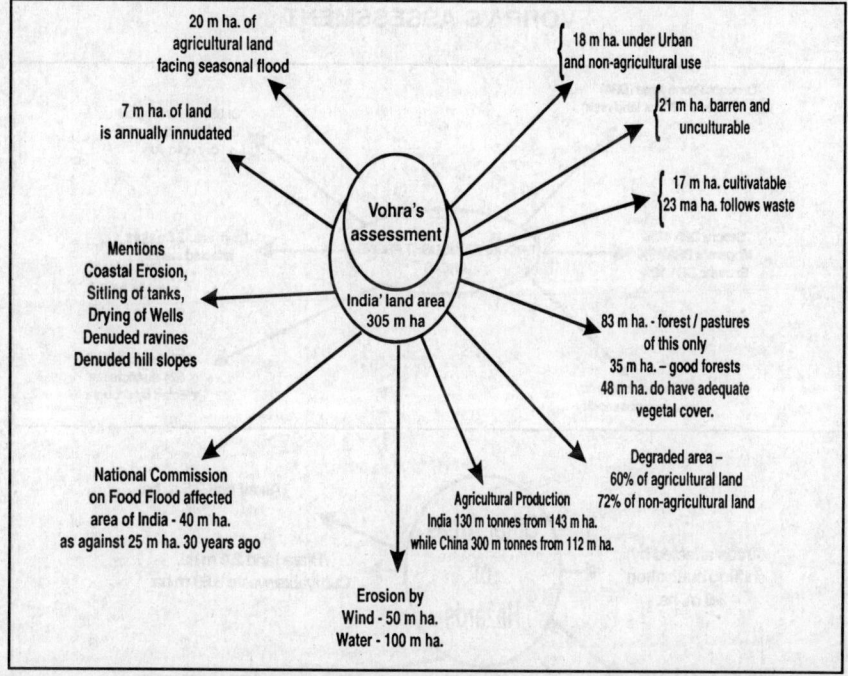

"It has been estimated that about 20 million ha. of good agricultural lands are subject to seasonal floods – the average area inundated annually being about 7 million ha."

"Denuded hill sides, ravines, water logged and saline lands drought stricken villages, silted tanks and drying wells are to be encountered almost everywhere. Vohra further says that":

- Flood ravage large areas year after year, even as the Rajasthan desert maintains its leeward creep.
- Erosion of coastal areas, particularly in Kerala is a major problem.
- A surprisingly large number of our planners, politicians, policy makers and economists, still believe that there is nothing very much wrong with the manner in what India managed to land resources.
- Vohra calls it "resource illiteracy" of our planners.
- Of the area of the country 305 million hectares, (i) 18 m ha. are under urban and non-agricultural use, (ii) 21 m ha. are barren and unculturable.
- Of the balance 305 – (18 + 21) or 266 m ha. (i) 17 m ha is cultivable waste, (ii) 23 m ha as follows/(iii) 83 m. ha. of forest, only 35 m ha. has good tree cover, the balance 48 m ha. do not have any vegetation.

VOHRA'S ASSESSMENT

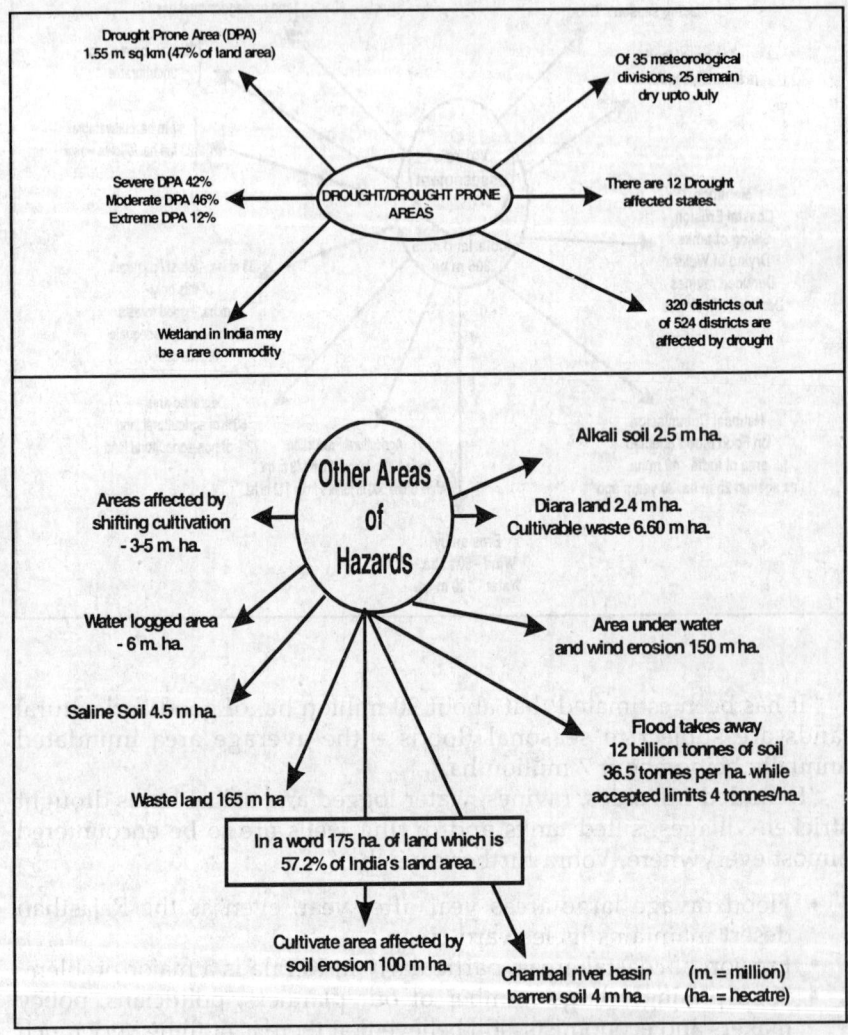

Vohra has drawn conclusion that fully one-third of our total relevant land area and three quarter of our total non-agricultural area is today lying practically useless.

The figures put forward by Vohra shows how sick our non-agricultural lands are. According to information released by the Ministry of Agriculture recently as many as 175 m ha. equivalent to 66% of the area are affected by degradation caused by serious soil erosion and water logging and salinity which, incidentally are the only two major ills that the land suffers from.

About flood Vohra says:

- "According to the National Commission on Floods, the area affected by annual floods today starts at around 40 million ha. as against 25 million ha. about 30 years ago. The second factors concern the growing menace of water logging and salinity in newly irrigated canal command areas."
- "Category-wise at least 61% of our agricultural lands and at least 72% of our non-agricultural lands are degraded to a greater or lesser degree."
- "No wonder we can barely manage to provide 130 million tonnes of food grains from 143 m ha. of agricultural land, while China produces significantly more than 300 m tonnes from mere 112 m ha. No wonder destitution and unemployment stalk the country."
- "It is estimated that the area seriously affected by wind erosion is around 50 million ha., while the area seriously affected by water erosion is about 100 m.ha."
- "It is sad but undeniable fact that our forest departments, have by and large, failed to protect over 70 odd million hectares of forest land against encroachers' fellings, and unauthorised denudation. This is because they do not have adequate legal power over outside Reserved Forest."

In Nepal the destruction of forests is taking place at such a rate that the country is likely to be all but totally denuded by the end of the century. We should be particularly concerned over this development because soil erosion in Nepal has a direct link with floods in the Ganga basin.

The gigantic task of reclosing 131 m ha. of bare land has to be accomplished.

Reclamation of 10 m odd hectares of land will need Rs. 15,000 crores.

Drought and Flood are Positive Threats

Several views on the subject are placed below which will show how horrendous is the land use situation in India. The figures presented by various offices may differ as exact figures are not available from a valuable source; but they definitely indicate a serious situation. Some assessment figures and observations are as follows:

- Dr. Hari Narain, Director, National Geophysical Research Institute, Hyderabad spoke in symposium in 1969.
  - "The droughts and floods have been the recurring problems of this country and will continue to be so for many years to come. Unfortunately, so far the scientific approach has been lacking to a great extent."

- Drought Prone area
1.55 million sq. km. or about 47% of the total area of India.
Of this severe D. P. area covers 42%.
Of this moderate D. P. area covers 46%.
Of this extreme D. P. area covers 12%.

[Source: **Proceeding of Symposium**
by Royal Aseema Geographical Society, 1979]

Drought, Soil Loss, Erosion, etc.

- India is losing 6000 million tonnes of top soil per annum through water erosion.
- Wetland in India may be a rare commodity in near future.
- The land use situation nullifies to utilize the benefit of having 58 major and medium rivers.
- There are more than 300 drought prone districts of India (Greenery, Agricultural land, Vegetation, Water bodies are affected).

[Source: **Ministry of Agriculture and**
Cooperation, 1982]

Assessment by other Agencies

Vohra (1978)
- Degraded areas –35 million ha.
- Ravine affected areas – 7369 million ha.

Ground Water

- India's ground water resource 418 billion cubic metres (cum)
- At present 100 billion are being exploited.
This leaves only 1528 billion cum of utilizable water resource of which 276 billion cum could only be harnessed.
- Planning Commission (1982) assessment is:
  - Area under water and wind erosion – 150 million hectare (mh)
  - Shifting Cultivation   –   3 mh
  - Water Logging   –   6 mh
  - Saline Soil   –   4.5 mh
  - Alkali Soil   –   2.5 mh
  - Diara Land   –   2.4 mh
  - Cultivable wasteland   –   6.60 mh
or 175 mh or 57.2% of India's land are

[Source: **Planning Commission, 1982]**

Degradation Sources Identified

• Exhausted due to over grazing; scarcity of fuelwood; Degradation of agricultural land; Damage of water bodies; Loss of humus due to burning; Repeated flood and soil erosion; Rapid rate of increase of human population; Rapid increase in livestock population; Loss of resource due to shifting cultivation.

Other Quantitative Assessment of Loss/degradation

• Waste land – 165 (million ha.) (1980) is more than half of geographical area.
• Flood takes away 12 billion tonnes of soil annually (or 36.5 tonnes per ha. while accepted limit is 4 tonnes/ha.).
• Cultivable area affected by soil erosion – 100 million ha.
• Barren hills, Soft friable, swaliks, unconsolidated ravines of Chambal covers 4 million hectare.
• Salines – Alkaline soil and water logged areas of 20 million hectare.
• Shifting cultivation affects 5 million hectares.

Dry and Degraded Environment

• Desert Zone	–	9,29,000 ha.
• Arid Zone	–	1,34,55,200 ha.
• Semi Arid Zone	–	6,68,3,000 ha.
• Sub-Humid Zone	–	11,00,00 ha.
• Humid Zone	–	2,11,000 ha.
Total		2,13,88,200 ha.

Land Degradation due to Erosion (in hectares)

• Water Erosion	–	90,00,000 ha
• Wind Erosion	–	50,00,000 ha
• Seasonal Flood	–	12,00,000 ha
• Inundated Annually	–	07,00,000 ha

[N. B. India's land area is 328 million ha.]

Degradation – FAO (1981) Assessment

• Scrub Forest (No trees)	–	5.378
• Open Forest (scattered trees)	–	5.393
• Forest subject to heavy biotic pressure	–	4.324
• Rate of degradation	–	0.8/year

LAND USE

Accelerated erosion :

Water erosion	– 90 million ha.
Wind erosion	– 50 million ha.
Seasonal flood	– 12 million ha.
Inundated annually	– 0.7 million ha.
Total land area	– 328 million ha.

Various Hazards : Drought prone area 155 m. sq.km.

- Ravine affected area—73-69 lakh ha.
- Degraded areas 35 million ha. (Vohra 1978)

FAO 1981 Degraded Forest 15.095 million ha.

- Scrub (no tree) – 5.378 m. ha.
- Open forest (scattered trees) – 5.393 m. ha.
- Forest, subjected to heavy – 4.324 m. ha. biotic pressure.
- Rate of degraded in 0.8 million ha. per year Khan 1987.

Various Hazards :

- Exhausted by over grazing.
- Scarcity of fuel wood and fodder.
- Degradation of agricultural land.
- Damage of water bodies.
- Deprived of humus due to constant burning.
- Repeated flood and soil erosion causes of degradation.
- Increase in population and clearance for agriculture.
- Increased live stock population.
- .... for fuelwood timber, shifting cultivation.

India's Horrendous Land Degradation Situation

Various arid zones:

Desert zone	–	929000 ha
Arid zone	–	13455200 ha
Semi-arid zone	–	6686000 ha
Sub humid zone	–	110000 ha
Humid zone	–	211000 ha

Desertification

Arid area 3200000 ha

(12% of country's area)

• Paper and paper pulp
  – free chlorine
  – sulphide, sulphurous acid
  – pental chlorophenol
  – dehydroabietic and
	• Starch and Sugar mills
	
	• Textile mills
	
	• Rayon Plants

• Bleached Kraft Mill
  – phenol, chlorophenol
  – sodium salts of unsaturated fatty acid
  – sodium isopirnarate   **Industrial waste and damage to aquatic life**
	• Fertilizer Plants
	
	• Petroleum refineries

• Breweries and Distilleries	• Dying industries
• Tannery work	

DRAINAGE AREA OF LARGE RIVERS (NEEDS LARGE SCALE AFFORESTATION AND SOIL CONSERVATION)

A realistic picture of the country's drainage area of several important rivers of India has been drawn to show that these vast catchment areas need sufficient effective vegetal cover and a vast environment oriented landuse planning.

Otherwise, there will be wide scale siltation of rivers (many have already been silted) resulting in spilling of water from rivers causing flood.

The vastness of the problem is highly significant and need mobilization of a large sum of money and human resources.

Statewise Inside India Data (area in sq. km.)

Uttar Pradesh	2,94,364 (34.2%)
Rajasthan	1,12,490 (13%)
Madhya Pradesh	1,98,962 (25.1%)
Bihar	1,43,961 (16.7%)
West Bengal	71,485 (8.3%)
Himachal Pradesh	4,317 (0.5%)
Punjab and Haryana	34,341 (4%)
Delhi	1,484 (0.2%)

Drainage of River Basin (area in sq. km.)

The Chambal basin

Madhya Pradesh	57948
Rajasthan	80670
Uttar Pradesh	850
Total	139468

The Jamuna basin (excluding Chambal basin)

Himachal Pradesh	4317
Haryana	34283
Punjab	58
Uttar Pradesh	73763
Madhya Pradesh	81030
Rajasthan	31820
Delhi	1484

The Ramganga	32493

Basin between Ton and Son

Madhya Pradesh	12328
Uttar Pradesh	10909
Bihar	532
Total	28569

Gomti, Ghaghara and other rivers between Jher

Uttar Pradesh	98127
Bihar	3000
Total	101127

The Son basin

Madhya Pradesh	47656
Uttar Pradesh	5952
Bihar	17651
Total	71259

The Gandak and other rivers

Bihar	47535
West Bengal	8795
Uttar Pradesh	968
Total	57298

The basin of right bank tributaries east of the Son

Bihar	61261
West Bengal	35843
Total	17104

The main Ganges

Uttar Pradesh	71302
Bihar	9182
West Bengal	26847
Total	107331

The Brahmaputra basin

Arunachal Pradesh	81424
Assam	70634
Nagaland	10803
Meghalaya	11667
West Bengal	12585
Total	187113

The basin of the Barak and other rivers

Assam (also Mizoram)	28890
Meghalaya	10775
Manipur	22347
Tripura	10453
Nagaland	5685
Total	78150

Basin of West flowing river between Kanyakumari and the Tapti

Tamil Nadu	4702
Kerala	35925
Mysore	25095
Goa	3610
Maharashtra	32573
Gujarat	9666
Daman, Dadra, etc.	546
Total	112117

The Tapi basin

Madhya Pradesh	9804
Maharashtra	51504
Gujarat	3837
Total	21674

The Luni and other rivers of Saurashtra and Kutch

Rajasthan	193392
Gujarat	128920
Diu	39
Total	321851

The Indus basin

Jammu and Kashmir	193762
Himachal Pradesh	51356
Punjab	50304
Haryana	9939
Rajasthan	1584
Chandigarh	114
Total	321289

[*Source:* **Report of Irrigation Commission, G.O.I. 1972**]

India Faces an Imminent Land-use Calamity and Disastrous Future

While this author was engaged in enumerating the extent and size of various natural calamities faced by India he found a proverbial similarity in the conception pattern of our planning objective with the famous Roman Empire 'Nero'. Nero, it is said, was playing his musical instrument in delight to build up a new city of Rome while entire city of Rome was ablaze said to have been set on by him. Just at that very ominous moment the T.V. flushed the devastating and hair rising news of the disaster of "Tsunami". It seems Nature God himself supported the author's apprehension that the Mother India was facing almost a similar crisis caused by extensive soil erosion, landslide, flood, drought, scarcity of drinking water, inadequacy of irrigation water, extension of desert and drought prone areas, denudation of forests, over-grazing, forest fire, pilferage of forest resources and other such problems; these factors have reduced the very foundation of the country to a fragile, dilapidated and brittle landmass which may give way any moment; on this unstable soil. Various development schemes are being formulated.

Over and above there is frequent threats of earthquake and the effect on El-Nino (drought) in Orissa and may be elsewhere also. These grim situations are being undermined and building up a prosperous modern India with borrowed money little realising that entire country might go under the Indian Ocean due to perilous factors mentioned above. Besides, utter poverty, illiteracy, unemployment, absence of adequate petrol/diesel/gas, etc. resource and lately of insurgency and border intrusion are burning factors to nullify the planner's building of rejuvenated India.

The author has listed, presenting dependable latest official statistics, the derogatory factors which may bring utter disaster to this country that looks horrendous. The planning without considering these factors while building up millions of superstructures on soft, dilapidated and cracked foundation seem to be oblivious of India's fragile socioeconomic and sociocultural backwardness reeling under wide scale poverty, illiteracy, health hazards and primitive social systems.

Swami Vivekananda's clarion call made a hundred year's back for a man-making and character building education has not yet struck a root. The millions of illiterate and poor masses have yet to develop a strong will to lift themselves up from the present state. We hardly realise (once again refer to Swami Vivekananda's say) that unless these people themselves rise to the occasion and want to improve their fate, no amount of money poured to their villages will improve their condition.

The author feels that India does not need at the moment multicuisine restaurants like the ones in Calcutta (like Red Kitchen, Lounge Alipore, China Bistro, etc.) that serve unusual mix of Chinese, Afghan, Thai, Tibetan and Italian cuisine or luxurious extravaganza of five or seven star hotels, multiplex, sophisticated and expensive fashion and beauty parade, film festivals dances of ill-clad dancing and gyrating girls, cheap Indian pop song, burning of precious commodities like petrol/diesel/gas in car-racing, luxurious travels, loss of precious human energy caused by strikes, bunds and even 5 day 24 hours cricket games (our neighbouring country Bhutan banned T.V till 1999), and expansion of liquor shops and the like.

The planners do not feel that Calcutta intelligentsia (connoisseur of Cultures they say) must not now dream of all night adda (rendezvous) or over 300 days a year and dream of the sights and sounds of the city in the grip of festive fever, scent of flavours of thousand savoury and sweet delight, roam about the whole night in puja pandels, indulge hours in fun and froth being oblivious of crimelords at the nations border, of local gangsters and other toughest challenges are posed by intrusion inside the country by foreigners. Our intelligentsia are environment conscious on paper and never feel that the need of the day is massive afforestation work, extensive mine, soil conservation engineering works and stoppage of pilferage of natural resources at any cost.

The New Colour of Peace is Green

Indian youth must learn from Wangari Maathai who won the Nobel peace prize for her very persistent and sincere campaign to protect Kenya's forest and distributing food to hungry constituents suffering from drought.

Maathai was in the countryside – just a hill from her childhood home – when told she had joined a club that includes Nelson Mandela, Kofi Annan and the Dalai Lama. "This is an overwhelming experience. It is unbelievable, it's the kind of thing you never hear in your life. I'm very flattered", she said, adding she thought she was selected as a symbol of the struggles against poverty and environmental degradation in Africa.

The prize comes after a nearly 30-year campaign to protect and restore the environment as well as defend human rights. "Many of the wars in Africa are fought over natural resources", Maathai, Kenya's deputy environment minister, said. "Ensuring they are not destroyed is a way of ensuring there's no conflict".

With 194 nominations, the committee had a broad field to choose from and could have conferred the prize of someone tied to a hot-button issue, like nuclear proliferation. Instead, the committee tied the environment to peace – a decision that widens the scope of what qualifies for the award.

"We're delighted the Nobel committee has put the green into peace", a Green peace spokesperson said.

Kenyan environmentalist Wangari Maathai won the Nobel prize for her fight against poverty by trying to save the continents shrinkage forests.

"It cannot get any better than this – may be in heaven", Maathai said of the prize. She wept with delight and celebrated by planting a tree in her home town of Nyeri in the shadow of Mount Kenya, Africa's second highest peak.

The award, the first Nobel given to an environmentalist, marks a new interpretation of 1895 will of Swedish philanthropist Alfred Nobel which set up the prizes.

Some critics said the green theme betrayed mainstream peacemakers, but the Norwegian Nobel Committee defended its decision.

"Peace on earth depends on our ability to secure our living environment", committee head Ole Danbolt Mjoes said. "We have emphasised the environment, democracy building and human rights and especially women's rights", he said of the prize. "We have added a new dimension to the concept of peace".

Maathai's Green Belt Movement, comprised mainly of women, says it has planted 30 million trees across Africa to combat creeping deforestation that often deepens poverty.

Mjoes said the movement also worked for family planning, nutrition and the fight against corruption in Kenya that has allowed the felling of vast tracts of forests.

Maathai, 64, who is Kenya's deputy environment minister, said her grassroots movement could be a pre-emptive strike to safeguard peace.

"Many wars in the world are actually fought over natural resources", she told NRK Norwegian radio. "In managing our resources ... we plant the seeds of peace, both now and in the future".

But some were unconvinced. "You don't give Nobel chemistry prize to a professor in economics", said Carl Hagen, leader of Norway's Opposition far-right Progress Party. "A peace prize should honour peace, not the environment".

"This prize could be positive in expanding the concept of security, but it could also mean a dilution of the prize, moving to far away from the original idea", said researcher Espen Barth Eide at the Norwegian Institute of International Affairs.

He had tipped the UN nuclear watchdog and its head, Mohamed El Baradei, for the award to reflect global fears that terrorists or rogue states might obtain nuclear arms.

But others argued that environmental degradation posed a huge long-term threat to life on earth.

"Understanding is growing throughout the world of the close links between environmental protection and global security", said Klaus Toepfer, head of the Nairobi based UN Environment Programme. He called Maathai "Africa's staunchest defender of the environment".

"It's a happy day for every blade grass in Africa", said South Africa anti-AIDS campaigner Zacke Achmat, who had been among the candidates for the prize.

Maathai is the 12th woman peace laureate since the first award was made in 1901. The last African laureate was UN Secretary General Kofi Annan, of Ghana, in 2001. The 2003 prize also went to a woman, Iranian human rights lawyer Sirin Ebadi.

When Wangari Maathai got word she had won the Nobel Peace Prize, she was busy with the work that earned her the award – campáigning to protect Kenya's forest and distributing food to hungry constituents suffering from drought.

India has to shun mega-dream issues like interlinking rivers as the project will create a unending problems like transmigration, relocation of human settlement, and take up long term social, economic and environmental issues.

This project may need Rs. 500,000 crores which is about the country's one-third Gross National Product (A. K. Mitra, The Telegraph, Nov '04) and the problems it creates will be thousand times greater than Narmada Valley Project (A. K. Mitra). This effort is sure to cause crores of peoples' nightmare. Entire system of natural water flood will be effected besides the movement of massive quantity of earth, rock, forests and millions of people.

The Narmada project is a small dimension of the anticipated problems and the execution of the dream is likely to engender more than a thousand times more formidable problems. The ecological challenges alone are mind boggling.

India is facing various divisiveness that are plaguing the nation, and political convúlsions to ensue in the wake of the first endeavour at interlinking rivers are altogether unpredictable and some actually fear a collapse of whatever little cohesion there is still left in the nation (A. K. Mitra).

The immense problem of rehabilitation of uprooted people, added by the problems of Jammu and Kashmir, Assam, Manipur, Nagaland, insurgent groups all over the country and many other issues like unemployment, petrol shortage, drought, flood are indicative of burying this mega dream for some years.

Is India a Dying Landscape? Terrible State of Land Threat

Land Situation

India covers 329 million hectares of geographical area. This huge land mass can be divided into seven physiographic units, namely the Himalayas, Gangetic plains, central highlands, peninsular plateau, eastern and western coastal belts and the Islands.

The great mountains mass of Himalayas forms an impenetrable barrier to the influence of cold winds from central Asia and gives the small continent the elements of tropical climate. The country receives rain from southwest monsoon from June to September, where as the northeast monsoon brings shower from October to December. The rainfall varies widely being less than 150 mm in western Rajasthan to more than 3000 mm in parts of Assam, Meghalaya, Kerala, coastal Karnataka and Konkan coast. The annual normal average is about 1345 mm The forest covers hardly 32 million ha. Soil erosion, drought, flood are an annual feature and desertification of major land area is at sight.

The annual degradation rate of land is as much as 2.5 million hectares (Murty 1994), computed waste land of the country is put at 165 million hectares (1990) which is more than half the geographical area of country. The flood carry away 36.5 m. tonnes per ha. while the acceptable limit is 4 tonnes per hectare.

About 100 million hectares of cultivable land is affected by soil erosion. Soil erosion due to floods is estimated at around 12000 tonnes annually. Of this, about 30 per cent goes to the seas, 10 per cent is deposited in reservoirs, while the rest 60 per cent goes all over the country.

Barren hills, soft friable swaliks unconsolidated ravines of Chambal which covers 4 million hectares, saline-alkaline soils and water logged area of 20 million hectares and 5 million hectares Jhum land depict a miserable landscape.

Water is the inevitable media where life was first created. "Water is the essence of earth and plants are essence of water" observes *(Chhandogya Upanishad)*. But water is still rare commodity.

Ground Water

Of 418 billion cubic meters of ground water about 100 billion cubic meters are being exploited at present. This leaves only 1528 billion cubic metres of ultimate utilizable water resource in this country of which 276 billion cubic meters could only be harnessed. The stupendous and most important task before the country is to harness the 1252 billion cubic meters of water.

There are more than three hundred drought prone districts in India. This condition has affected agricultural sector adversely, destroyed greenery, affected agriculture, eroded soil, depleted vegetation, dried-up waterbodies and finally caused desertification.

This situation clearly indicates a positive threat and a grim picture for development of agriculture in India. Whatever little land, is left with, is threatened with depletion. Unless the management of land, water energy, agriculture, irrigation, soil and water conservation, forestry, etc. are integrated under one umbrella for a firm policy and action plan there is little hope of saving 80% of the wetlands other than the reservoirs and big rivers (Murty 1994). But the situation is still worse.

India is endowed with a good number of rivers but the distribution of availability of water is not uniform. The annual precipitation, including snowfall is estimated as about 5000 billion cubic meters; of the rainfall of 3000 billion cubic meters, 1853 billion cubic meters is average annual natural flow. Of this again only 1110 billion cubic metres could be put to beneficial uses by conventional means (Murty 1994).

Though India is blessed with great rivers and their basins, the major and medium rivers are 12 and 46 account for 92 per cent of the total run off and minor rivers (innumerable) have the balance 8 per cent runoff, the land use situation nullifies this benefit and the overall situation is a serious concern to this country.

Wetlands in India may in near future become a rare commodity. In the last 50 years, most accessible areas have been converted into settlements and paddy fields and other agricultural land. Modern drainage techniques have created many potential agricultural land. The population pressure also has forced to rehabilitate such areas into human habitation. The dense human population supplemented by a heavy cattle population. depleted substantial floral and faunal resources of these wetlands which may now be marked as a threatened or lost wet landscape.

In the precincts of urban areas most of swampy grounds have been reclaimed for construction of settlements. This is a picture common all over the country.

This 175.00 m.h. is about 57.2 per cent of land area which is 328 m.h. and has been quantified in a report of Ministry of Agriculture and cooperation (1982) which further states that India is losing 6000 million tonnes of top soil per annum through water erosion and that these represented in terms of major nutrients N, P, K alone an annual loss of Rs. 700 crores. The National Commission on flood (1980) puts the losses on account of floods in 1976, 1977 and 1978 to Rs. 889/- crores, Rs.1200- crores and Rs.1091/- respectively; according to the same report the total area subject to periodic floods which was estimated to 20 million hectares in 1971 now stands at the level of 40 million hectares – an increase of 100 per cent in 10 years.

The Himalayan ecosystem has considerably deteriorated resulting in floods in the Indo-Gangetic plains causing heavy damage to property and crops and untold misery of the people. Seepage of water to subsoil level is minimal due to heavy surface runoff which cause erosion and depletion of ground water resources.

Environmental Crisis in India

The extent of forest cover is a good indicator of the health of a land. Large scale deforestation in recent decades has reduced the sensitive catchment areas in the Himalayas and other hilly areas extremely vulnerable to soil erosion. India can achieve ecological security only increasing the vegetal cover to tackle the problem serious degradation of land which are (source – Planning Commission 1982; figures in bracket indicates million hectares) – Serious Water and Wind Erosion (150 m.h.), Shifting Cultivation (3 m.h.), Water Logging (6 m.h.), Saline Soil (4.5 m.h.), Alkali Soil (2.5 m.h.), Diara Land (2.4 m.h., other cultivable waste land fit for reclamation (6.60 m.h.) or in total 175 million hectares (or 57.2 per cent of India's land area) of land is reeling under annual flood, erosion and drought (details in Chapter 1). The forest cover according to National Remote Sensing Agency (NRSA) 1984 has dwindled down to about 19.5% from much quoted 22.7 per cent; the density has decreased (areas having only 10 per cent density) from 55.18 m.h. to 46.08 m.h. in eight years. There has been depletion of forests at a rate of about 1.5 m.h. per annum. The NRSA calculated the area of good forest to about 33 m.h. in 1984. Of late the Forest Survey of India (FSI) has reconciled the deviation of figure of both FSI and NRSA and arrived at a consensus figure. The cover is now put at 19.528 per cent, the dense and open forest comprising about 50 per cent each. (Map 2 showing original cover).

Drought and Flood: The Cruelest Enemy for India's Progress

A day's rainfall causes flood and resources damage and a day's scorching sun causes drought and crop failure, such is the precarious condition of land use in India. The situation is touching a point of no return.

The facts and figures put forward relating Threats have also been upheld in the media report, published in the daily "The Telegraph", July 25, 2002; few relevant portions of which are reproduced below:

"It is time of year when India's famed diversity becomes stark-while some states are hit by floods, others are on the threshold of drought. Every year, myriad of rituals are held in July to apprise the rain god. Marriages of frogs and donkeys are organized, animals are sacrificed, yagnas performed, children abducted, desperate women parade naked on parched fields and nomadic eunuchs are assigned to do the rain dance. Annual rituals to ensure a modicum of respite from the drought."

Of the 35 meteorological divisions in the country, 25 remained dry till last week, with the cumulative rainfall from June 1 to mid-July falling short in 20 divisions. The "variation cycle" is rare, but been recorded in 1992 and 1995.

The scanty rainfall has impacted the power scenario in most states, especially Andhra Pradesh and Orissa. There has been a steep rise in power consumption in the cities, which has further worsened the situation. The drought, which will lead to depressed rural demand and income will also threaten the bottom-lines of corporate India. But instead of implementing a comprehensive contingency plan, the Centre has decided to wait and watch until July 31.

"The Union Minister for Agriculture, Ajit Singh, had a marathon meeting with relief commissioners of 12 drought-affected states yesterday. These included Andhra Pradesh, Chattisgarh, Haryana, Himachal Pradesh, Karnataka, Madhya Pradesh, Maharashtra, Orissa, Punjab, Rajasthan, Tamil Nadu, and Uttar Pradesh. But besides declaring that this is country's worst drought in 12 years (320 to 524 districts are affected and unfolding a four-fold 'preliminary package', what they came up with) could have been done weeks ago – seed procurement methods, operational conditions of irrigational facilities, rescheduling of loans to farmers, drought resistant crop varieties and the like.

The situation is hardly different in the flood-affected areas. Embankments are seldom reinforced before the rainy season. Standby shelters are always an afterthought. States like Bihar and Assam are prone to floods every year; yet the routine circus of "surveying" the vast, flooded plains precedes any concrete action plan. Closer home, anti-erosion work in flood-prone Malda was taken up well after the onset of the monsoons. While the irrigation and finance departments played pass-the-buck, boulders meant to fortify the embankments are rapidly being washed away by the rising waters of Ganges. Those affected are never shifted to higher ground when the first alarm is sounded. Like in previous years, they continue to live in dread waiting for the water to come swirling in one night."

"The country's size is usually cited as an impediment to farming disaster management strategies that are common in southeast Asia, especially in Indonesia and Malaysia which share our climatic pattern. The administrations in these countries are ready with artificial rain at the first sign of drought. In India, states like Sikkim and Maharashtra have exhibited a more pragmatic response to natural disasters. Take for example the landslides in hill states over the past fortnight. While the administration in Uttaranchal's Rudraprayag and Chamoli districts and Assam's Barail ranges looked helplessly as villages turned to rubble, in west Sikkim residents were evacuated well in time from remote hamlets in Dentam, Richenpong, Burmoik, Soreng, Daramdin, Yoksome and Tashiding. Of the 10 flood-affected districts in Assam, rail and road links to Dhemaji, which is only the first wave of floods, but the toll has already risen to seven, while five lakh people have become homeless and the world's largest river island, Majuli, lies half-submerged."

"We are verifying the facts. But people believe the Lord Ram's father-in-law, King Janak, and his wife ploughed the fields at night naked when their kingdom faced a drought and forced the heavens to open up. Similar rituals have been reportedly performed earlier also", said a state official.

The delay in the monsoon has sparked panic throughout Uttar Pradesh, one of the biggest food grain producing states in the country.

"Though drought has formally not been declared in the affected regions, the situation is alarming", state agriculture minister Hukum Singh said.

The state has registered a 25-30 per cent loss in kharif crop, mainly pulses and paddy, Singh said. "We have already arranged for the distribution of late-variety seeds of paddy and pulses to affected farmers. The delayed monsoon might also affect the rabi crop".

Technically, drought is declared only after a 50 per cent loss of crop, but a panicky government has decided to suspend tax collection in affected areas. As many as 58 to 70 districts are on the verge of being declared drought-hit.

"The situation looks grim this time. Though eastern Uttar Pradesh has received some rain, it is much below normal. The west and central regions are almost dry", Lucknow meteorological director K. Verma said.

While rainfall in the Bundelkhand region has been 65 per cent below normal, in western Uttar Pradesh it is 65 per cent below normal and in central Uttar Pradesh 51 per cent.

The government has directed district magistrates to send the situation report from their respective regions, but only 10 of 70 districts have responded.

"The prolonged dry spell has led to an alarming decline in underground water. There is also a power crisis in the state. While water in canals has dried, there is little power to run tube-wells. Last week, the total demand for power was 6052 mw but only 5052 mw was available". It is from paper quotation.

[*The Telegraph*, 19.7.2002]

This is the devastating situation that India faces today; and the land use situation is getting worse day-by-day.

Floral and Faunal Resources — Face a Crisis

Gone are the days when India's floral and faunal resources were considered very rich. In the past sixty years biotal scenario has undergone a drastic change due to undercontrolled biotic factors; some abiotic factors have also aggravated the situation. The Department of Forests have been in existence since the sixties of the nineteenth century and on its recommendations Wildlife Protection Acts have been enacted for conservation of 'Birds', 'Elephants' and 'Rhinoceros'. More recently the Wildlife (Protection) Act of 1972 for protection of all biotal species whose existence is threatened have been enacted. The Act takes care to protect various categories of flora and fauna in consideration of their present status. In the various management plans drawn up for Forest Divisions effort were made to preserve the threatened tree species; but little care was taken for the preservation of herb and shrub flora which form the bulk of medicinal plant material.

The author have made an endeavour to bring to light the various adverse factors responsible for depletion of flora and fauna especially the ground flora.

However, much we boast of our rich floral and faunal wealth especially the vast resources of medicinal plants, the fact remains that such resources have suffered a severe setback during the last 60 years. Many species may not yet have become extinct but surely their number has gone down to an alarming level. India is now limping into a new millennium with a much depleted floral and faunal resources, caused by uncontrolled increase in human and cattle population, rampant grazing, forest fire, acute soil erosion, and ruthless exploitation of biotal resources.

Medicinal plants are mostly herbs, some shrubs and climbers, and a few trees. These occur mostly in forest areas, some in scrublands, waste lands and in marginal lands. Forestry practices in upgrading mixed miscellaneous forests and forests comprising less valuable inferior species to plantations of species of high commercial value have depleted ground vegetation, resulting in reduction of biodiversity.

Several settlements, diagrams and maps have been presented in the books (refer bibliography) to provide a realistic picture of the types of impact on vegetation. These data bring out the linkage between population growth and the qualitative and quantitative parameters of forest resources in the country and highlight the colossal degradation of forest cover (density per unit area).

India's evergreen, moist deciduous and temperate forests are the primary home of medicinal plants apart from the country side in general. However, the dry-deciduous forests and the dry zones of the country also bear stray herbaceous plants (including medicinal plants) to some extent. India may not for long hold inexhaustible floral resources, especially of medicinal plants, and thorough quantitative study will testify to his statement which has not been so far. The Forest Survey of India has made a very useful assessment of the density, frequency, abundance and volume assessment of important tree species all over the country. Some states also have such figures for trees worked out for their forests. But data on herbs, shrubs and climbers are hardly available.

In this context one has to evaluate the effectiveness of the conservation efforts of the Government, whether, these steps are adequate to maintain the ecological balance, biodiversity and quality of environment by controlling all the adverse factors. There is general awareness today that increasing population, rampant poverty, and inadequate institutional framework are the underlying causes of forest depletion and degradation.

Complicated Study

Any study on the status and availability of medicinal plant resources requires thorough qualitative and quantitative ground survey of vegetation cover, especially of herbs and shrubs. Besides, the impending pressure on such resources posed by a rapid growth of human and cattle population in a situation of poverty and faulty land use have to be taken into consideration. It is a multifaceted issue which involves a thorough knowledge of systematic botany, chemistry and biochemistry (phyto), pharmacology and of the socioeconomic disciplines.

The tribal people and fringe forest dwellers have been utilising forest plants since time immemorial. Rev. P.O. Bodding's book entitled 'Studies/Medicine Connected Folklore (1986)' is a valuable work on Ethnobotany and Ethnobiology that have anchored deeply among the tribal and rural masses of this country.

Plants are considered to be medicinal if they possess pharmacological activities of possible therapeutic use. In simple language it may be said that plants which are useful to human beings or benefit them are medicinal plants. According to the later definition most plants have some medicinal value. Besides, human knowledge may have failed to discover medicinal properties in some plants which cannot be discarded outright as having no medicinal value. One cannot pinpoint the actual number of plant species in India.

Diversity

The diversity of floral species in India is due to the occurrence of 16 different agroclimatic zones, about 10 broad phytogeographical divisions, more than 25 biotic provisions, 450 biomes (habitats of specific species) and 200 forest types besides, the region of Eastern Nepal, India (Sikkim) and Bhutan, being the confluence of biogeographic elements, viz. Malayan, Japanese, Indochinese, Ethiopian, Mediterranean, Oriental and Palearctic, contributes to diverse biota.

R. P. Chaudhuri in his work 'Herbal Drug Industry' mentions 130 chemical substances extracted from 100 species of higher plants now being used in medicines. A UNDP report (1994) mentions that medicinal plants obtained from developing countries may be valued at more than Rs. 1,00,000 crores. Innumerable plant species in India have yet to be brought under proper research to assess their medicinal value.

Several Indian research centres have started to seek patentable genes, using the latest genetic engineering techniques. More than 400 useful plants have been identified so far. (Tropical Botanical Garden and Research Institute – TBGRI).

Readers may refer to the author's book on "Biodiversity" in this regard. The Himalayan regions – Western, Central and Eastern have tridimensional facets (altitudinal, latitudinal and Spatial) which cover various climatic zones (tropical, subtropical, subtemperate, temperature and alpine). There is a rich rain forest belt in East Himalayas from Arunachal/Assam to Tripura across Manipur, Mizoram, Meghalaya and Nagaland that have more than 45 per cent of the plants of India. The

Indogangetic and Brahmaputra plains have a poor assemblage of plant species due to intensive cultivation of agricultural crop. The dry zone of Rajasthan, Punjab, Gujarat, Madhya Pradesh, Andhra Pradesh, Maharashtra, Karnataka and Tamil Nadu, Southwestern Karnataka and the Chotanagpur hills in Singhbhum, Mayurbhanj and Sundergarh which have a rich assemblage of flora. But shifting cultivation in all the states of Eastern India and some areas of central and western India has depleted flora and fauna over at least half the forested area in the plains and hill all over the country. Grazing in the forest has depleted herb, shrub and regeneration of flora in the ground beside disturbing the ground biota.

India's Critical Forest and Landuse Scenario at a Glance

Population increase is linked with forest resource shrinkage

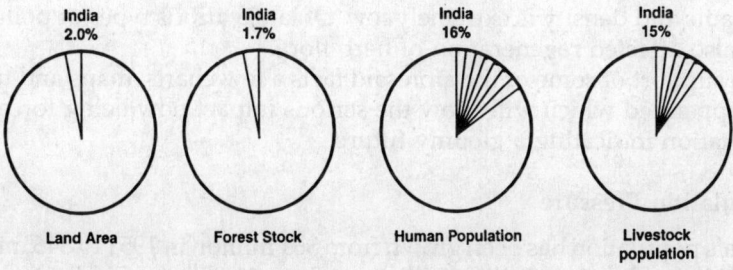

The serious land use situation is reflected in the above diagrams. With a sizeable share of human population (16%) and livestock population (15%) in the world. India has only 1.7% of the total area under forest stock in the world.

The resources of land and forest area are not commensurate with human and livestock population.

Factors Causing Depletion of Vegetation

Although India has a rich floral and faunal resource having a large number of genera and species, several adverse factors have threatened their existence which are:

– repeated heavy grazing by a vast population of domestic cattle.
– annual forest fire killing ground herbs and having derogatory effect on soil.
– shifting cultivation (Jhum) in the hills.
– ruthless exploitation of plant parts eradicating the mother plant materials.
– expansion of human population and mushrooming settlements.

- extension of cultivation.
- heavy erosion of land.
  • repeated flood and drought.
  • excessive deforestation.

Some of these factors have assumed a serious dimension damaging habitat, reducing effective areas of ground vegetation, density and frequency. City or town outskirts, village locations and peripheries and marginal lands have now been encroached upon or utilised for settlements and cultivation thereby reducing density of herb flora. In forest areas repeated fire and rampant grazing have forced only a few tough resistant species to survive the onslaught and by this process eradicating the softer species (Lantana, Eupatorium, Mikania, Ipomoea, Parthenium, etc. predominate, while Dioscorea, Rauwolfia, Podophyllum, Swertia, Aconitum, etc. alarmingly get eradicated). Even though some medicinal plants escape eradication, their occurrence is sporadic and density is extremely low. Of late, various types of pollution has also affected regeneration of herb flora.

In support of some of the aforesaid facts a few charts, maps and tables are appended which will show the serious impact now being forced on vegetation indicating a gloomy future.

Population Pressure

India's population has been grown from 361 million in 1951 to 846 million in 1991: and has crossed one billion mark on 11.05.2000 (Table 1).

Table 1: **Population : Size, growth and distribution pattern in India 1951-91**

Category	Population millions		Decadal growth rate of population (in percentage)				Ratio of 1991 to 1995 population
	1951	1991	1951-61	1961-71	1971-81	1981-91	
Total	361	845	21.5	24.8	24.7	23.9	234
Rural	299	629	20.5	21.9	19.3	20.0	210
Urban	62	218	26.4	38.2	46.1	36.5	349
Class-I Town	28	139	44.9	53.5	54.4	46.9	504
Million + Cities	12	71	54.1	53.7	51.3	67.8	601

Pressure on forests, pasture, agricultural land coupled with rural people's poverty and resulting migration to urban and industrial areas may be seen from the (Table 2).

So the vegetation is depleted, especially the herbs, which form bulk plant resources for medicine.

Pressure of Livestock Population

Table 2: India's Livestock population : 1951 to 1992 (in million)

Category of Livestock	Decadal population of Livestock (in million)				
	1951	1961	1972	1982	1992
Cattle	153.30	175.50	178.30	198.50	204.50
Buffaloes	43.30	51.20	57.40	69.80	83.50
Sheep	38.40	40.00	40.00	48.80	50.80
Goats	47.10	60.90	67.50	95.30	115.30
Horse/Ponies	1.50	1.30	0.90	0.90	0.80
Pigs	4.42	5.20	6.90	10.10	12.80
Camels	0.60	0.90	1.10	1.10	1.10
Others	1.30	1.20	1.10	1.30	1.30
Total	292.92	336.20	353.20	419.80	470.10
Standard Unit	228.00	259.70	270.50	304.50	336.00

Vast increase in sheep and goat population is a bad indication in forest conservation. The carrying capacity of India's Forest cover for grazing is estimated at only 31 million live stock, far less the number of livestock actually grazing in forested tract (Simon - 1986).

This situation will deteriorate as the demand for milk, egg, meat will go high with the increase in urban population and there will be further increase in livestock population.

Recession of Forest Cover

Forest cover of India has declined from 72 million hectares in 1951 to around 63 million hectares in 1997. A histographic representation for about 50 years is presented below, which shows decline in per capita forest area.

However, in the Table 6, the situation of forest cover and forest area per capita have been depicted which speak for themselves.

Situation of Fuelwood and Industrial Wood

In India firewood is still the primary source of energy for cooking. At present 78 per cent of rural household and 30 per cent of urban household (National Survey Organisation, 1997) use firewood as the source of energy for cooking.

Per capita available of forest cover in India

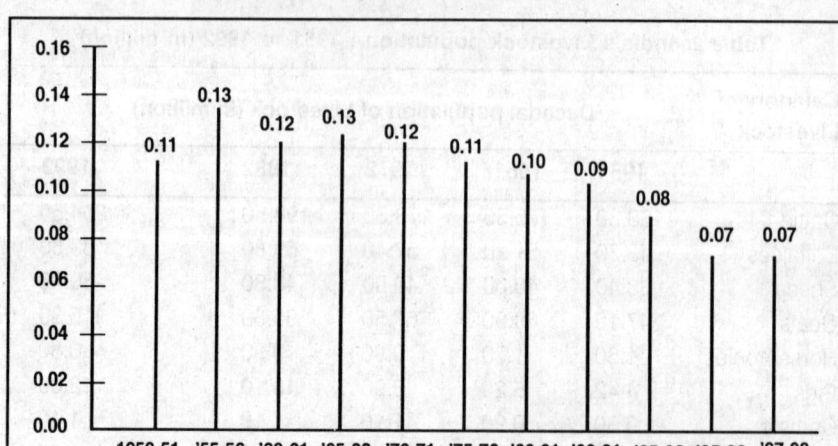

According to Tata Energy Research Institute, New Delhi, 2000 estimated annual consumption of firewood in India is around 325 million cubic meters, much above the carrying capacity and have presented a table (Table 3) and a Chart trend as follows:

Table 3: **Availability of fuelwood and industrial wood 1980-1994**

Year	Fuelwood Production	Industrial wood production	Fuelwood availability	Industrial wood availability
	Million m³	Million m³	Million m³	Million m³
1980	278.20	818.70	41.00	122.00
1984	276.30	651.90	37.80	89.30
1986	277.60	590.00	36.40	77.50
1988	279.90	450.10	35.20	56.70
1990	280.10	413.40	33.80	49.90
1992	280.60	320.70	32.50	37.10
1994	281.40	322.30	31.50	35.80

"Overall production of fuelwood in India has been stagnant over the period 1980-1994, increasing insignificantly from 275 to 281 million cubic meters. Despite this slight increase in total fuelwood production, per capita availability of fuelwood has been consistently declining (Table 4). In 1980 each person had 0.041 Cu.m. of fuelwood available by 1994 this figure declined to 0.031 Cu.m."

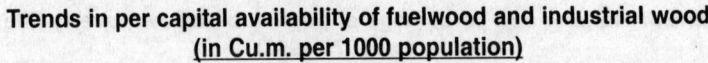

**Trends in per capital availability of fuelwood and industrial wood
(in Cu.m. per 1000 population)**

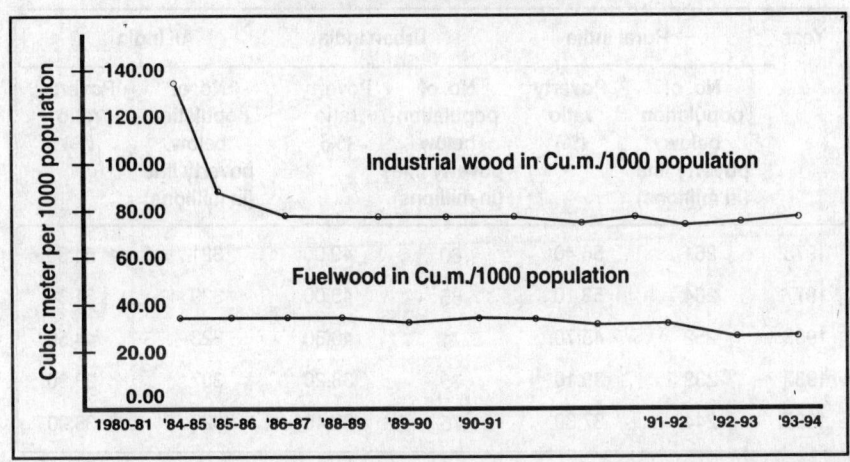

Likewise, industrial fuel production in India has consistently decline throughout the 1980-1994 period. Total production had declined from 819 million cubic meters in 1980 to only 322 million cubic meters in 1994. In this period per capita availability declined from 0.122 cubic meter to 0.036 cubic meter.

Table 4: **Over exploitation of forests**

Fuelwood Situation		Timberwood Situation	
Sustainable Capacity of Forests (in million m³)	Current Withdrawal from forests (in million m³)	Production capacity (in million m³)	Annual demand (in million m³)
48	235	12	28

[*Source :* UNFPA - 2000]

Poverty and Forest Linkages

Population and Forests : A report of India, Published by UNFPA (2000): On population and forests, quotes a figure from Economic Survey, 1998-99, which is reproduced (Table 5).

"This reveals a downward trend in both rural and urban poverty. India's anti-poverty and employment generation programmes - along with economic growth in recent years seem to have helped in reducing the overall incidence of poverty from 55 percent to the total population

Table 5: **Number and percentage of population below poverty line**
(in India, 1973-1993)

Year	Rural India		Urban India		All India	
	No. of population below poverty line (in millions)	Poverty ratio (%)	No. of population below poverty line (in millions)	Poverty ratio (%)	No. of population below poverty line (in millions)	Poverty ratio (%)
1973	261	56.40	60	49.00	321	54.90
1977	264	53.10	65	45.00	329	51.30
1983	252	45.70	71	40.80	323	44.50
1987	232	39.10	75	38.20	307	38.90
1993	244	37.30	76	32.40	320	36.00

in 1973 to 36 percent in 1993. However, the absolute number of persons below the poverty line has remained more or less the same in the last two decades i.e., around 320 millions, basically because of the pace of population growth during the period."

Pressure on natural resources like forests would reduce appreciably through generating employment opportunities in agriculture and other fields by poverty alleviation. Vast forest are easily accessible to the poor forest fringe dwellers for collection of fuelwood, fodder, food and minor forest produce.

That the growing population and heavy livestock population are linked to forest depletion and degradation resulting in biodiversity loss, which are clear from the fact presented in various tables.

India's forest resources have been declining for decades, while population has continued to increase. A Herculean effort on both population programmes and more holistic forest conservation schemes are needed now more than ever.

Forest Survey of India regularly monitor forest cover situation of the country. The position as in 1997 is presented in (Table 6). It shows the country's forest area to be 19 percent of total land-area. Actual situation of ground flora (obviously of medicinal herbs/shrubs) cannot be ascertained from this figure. Open forests have hardly a good cover of herbs/shrubs/climbers. Close forests (dense forests) which are also grazed and fire damage have caused in many areas a thin cover of medicinal herbs/shrubs. While inventory of density and frequency of tree crop has been done by Forest Survey of India but no such assessment has been done in respect of herbs/shrubs.

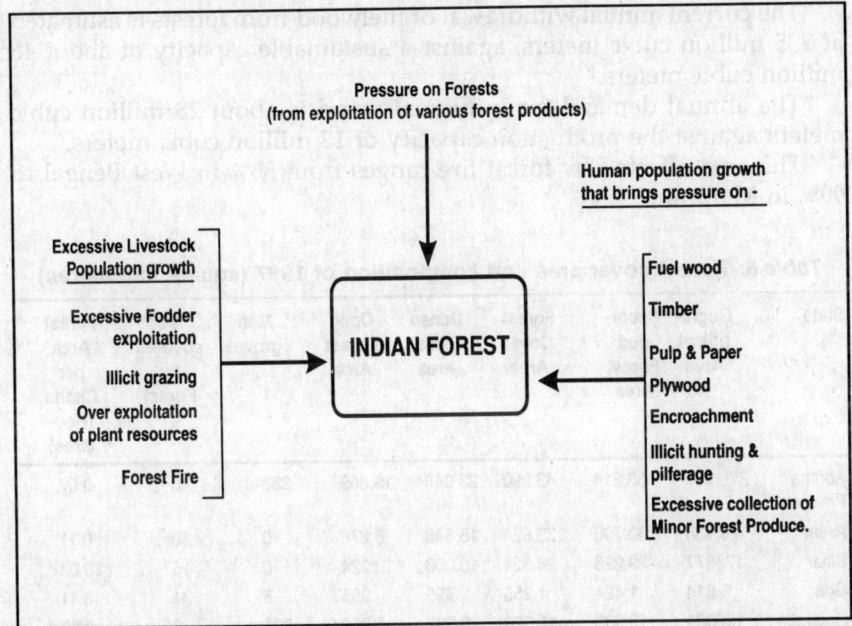

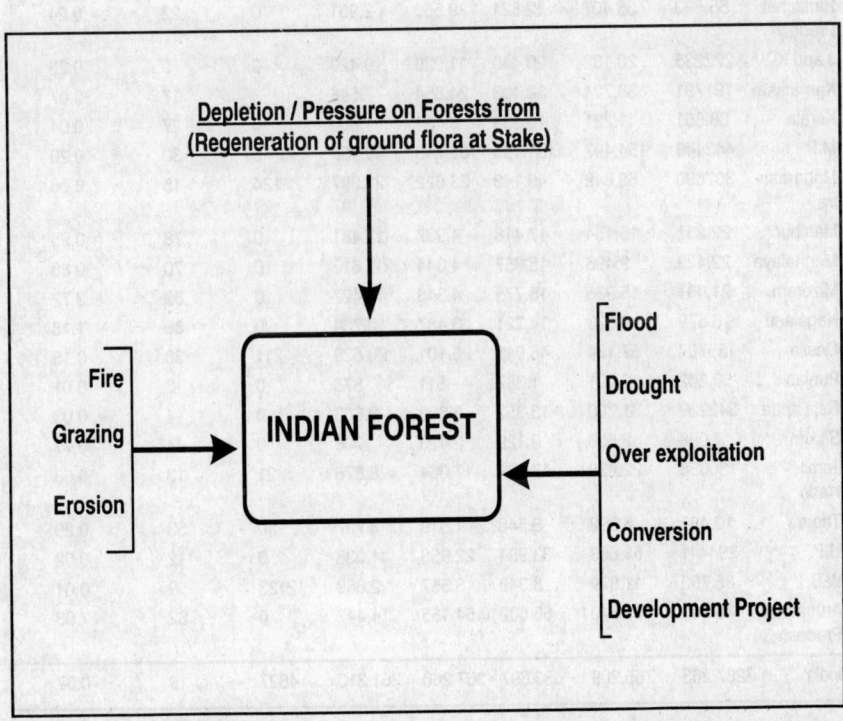

"The current annual withdrawal of fuelwood from forests is estimated at 235 million cubic meters, against a sustainable capacity of about 48 million cubic meters."

"The annual demand for industrial wood is about 28 million cubic meters against the production capacity of 12 million cubic meters."

"The area affected by forest fire ranges from 33% in West Bengal to 99% in Manipur."

Table 6: **Forest cover area and composition of 1997 (square kilometres)**

State	Gegraphical Area	Recorded Forest Area	Forest Cover Area	Dense Forest Area	Open Forest Area	Mangroves	% of covered by Forest	Forest Area per Capita (Hectares)
Andhra Pradesh	275.068	63.814	43.290	23.048	19.859	383	16	0.07
Assam	78.438	30.708	23.824	15.548	8.276	0	30	0.11
Bihar	173877	29.226	26.524	13.000	13.224	0	15	0.03
Goa	3.814	1.424	1.255	995	255	5	34	0.11
Gujarat	196024	19.393	12.578	6.337	5.250	991	6	0.03
Haryana	44212	1.673	604	370	234	0	1	0.00
Himachal Pradesh	55.673	35.407	12.521	9.560	2.961	0	23	0.24
J and K	222235	20.182	20.440	11.020	9.420	0	9	0.26
Karnataka	191791	38.724	32.400	24.854	7.546	3	17	0.07
Kerala	38.861	11.221	10.334	8.454	1.880	0	27	0.04
M.P.	443446	154.497	131.195	82.745	48.450	0	30	0.20
Maharashtra	307690	63.842	46.143	23.622	22.397	124	15	0.06
Manipur	22.237	15.154	17.418	4.937	12.481	0	78	0.95
Meghalaya	22.429	9.496	15.657	4.044	11.613	0	70	0.88
Mizoram	21.081	15.935	18.775	4.348	14.427	0	89	2.72
Nagaland	16.579	8.629	14.221	3.487	10.734	0	86	1.18
Orissa	155707	57.184	46.941	26.101	20.629	211	30	0.15
Punjab	50.362	2.901	1.387	511	876	0	3	0.01
Rajasthan	342239	31.700	13.353	3.690	9.663	0	4	0.03
Sikkim	7.096	2.650	3.129	2.423	706	0	44	0.77
Tamil Nadu	130058	22.628	17.064	17.064	8.676	21	13	0.03
Tripura	10.486	6.292	5.546	1.819	3.737	0	53	0.20
U.P.	294411	51.663	33.994	22.958	11.036	0	12	0.02
W.B.	88.752	11.879	8.349	3.557	2.669	2123	9	0.01
Arunachal Pradesh	83.743	51.540	68.602	54.155	14.447	0	82	7.93
India	3287.263	765.209	633.397	367.260	261.310	4827	19	0.07

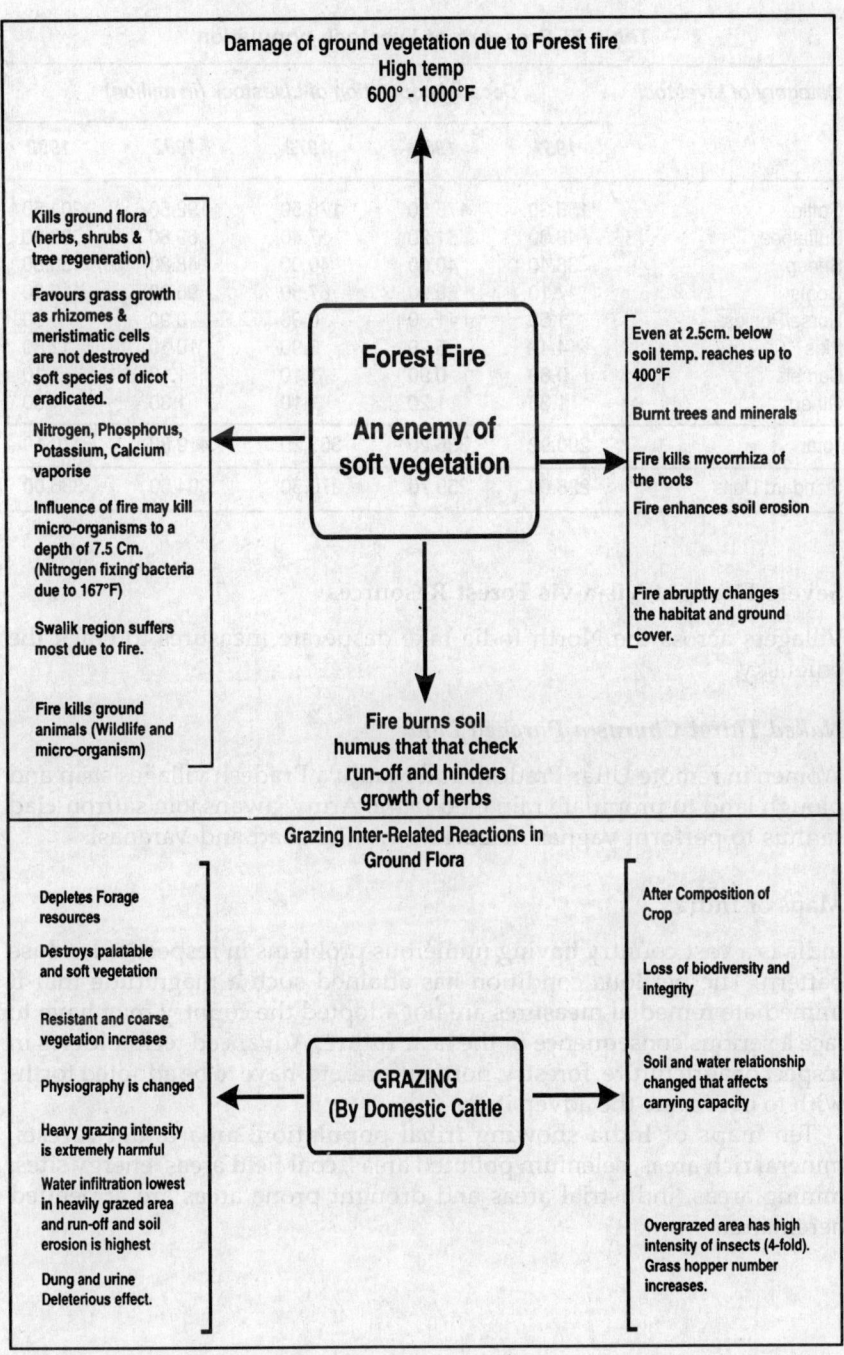

Damage of ground vegetation due to Forest fire
High temp
600° - 1000°F

Kills ground flora
(herbs, shrubs &
tree regeneration)

Favours grass growth
as rhizomes &
meristimatic cells
are not destroyed
soft species of dicot
eradicated.

Nitrogen, Phosphorus,
Potassium, Calcium
vaporise

Influence of fire may kill
micro-organisms to a
depth of 7.5 Cm.
(Nitrogen fixing bacteria
due to 167°F)

Swalik region suffers
most due to fire.

Fire kills ground
animals (Wildlife and
micro-organism)

Forest Fire

**An enemy of
soft vegetation**

Even at 2.5cm. below
soil temp. reaches up to
400°F

Burnt trees and minerals

Fire kills mycorrhiza of
the roots

Fire enhances soil erosion

Fire abruptly changes
the habitat and ground
cover.

Fire burns soil
humus that that check
run-off and hinders
growth of herbs

Grazing Inter-Related Reactions in
Ground Flora

Depletes Forage
resources

Destroys palatable
and soft vegetation

Resistant and coarse
vegetation increases

Physiography is changed

Heavy grazing intensity
is extremely harmful

Water infiltration lowest
in heavily grazed area
and run-off and soil
erosion is highest

Dung and urine
Deleterious effect.

**GRAZING
(By Domestic Cattle**

After Composition of
Crop

Loss of biodiversity and
integrity

Soil and water relationship
changed that affects
carrying capacity

Overgrazed area has high
intensity of insects (4-fold).
Grass hopper number
increases.

Table 7: Pressure of livestock population

Category of Livestock	Decadal population of Livestock (in million)				
	1951	1961	1972	1982	1992
Cattle	155.30	175.50	178.50	192.50	204.50
Buffaloes	43.30	51.20	57.40	69.80	83.50
Sheep	38.40	40.00	40.00	48.80	50.80
Goats	47.10	60.90	67.50	95.30	115.30
Horse/Ponies	1.50	1.30	0.90	0.90	0.80
Pigs	4.42	5.20	6.90	10.10	12.80
Camels	0.60	0.90	1.10	1.10	1.10
Others	1.30	1.20	1.10	1.30	1.30
Total	290.92	336.20	353.20	419.80	470.10
Standard Units	228.00	259.70	270.50	304.50	336.00

Severe Drought Vis-a-vis Forest Resources

Villagers across the North India take desperate measures to crack the cruel sky.

Naked Thirst Churns a Parched Land

Women in remote Uttar Pradesh and Madhya Pradesh villages strip and plough land to propitiate rain God *Indra*. Army jawans join saffron-clad sadhus to perform yagnas in Lucknow, Allahabad and Varanasi.

Maps of India

India is a vast country having numerous problems in respect of landuse pattern. The perilous condition has attained such a magnitude that if immediate remedial measures are not adopted the country may have to face a serious consequence in the near future. Advanced technologies in respect of agriculture, forestry, horticulture, etc. have to be adopted forthwith to overcome the adversities.

Ten maps of India showing tribal population, area under forests, mineral rich areas, selenium polluted areas, coal field areas, energy sites, mining areas, industrial areas and drought prone areas are appended hereinafter.

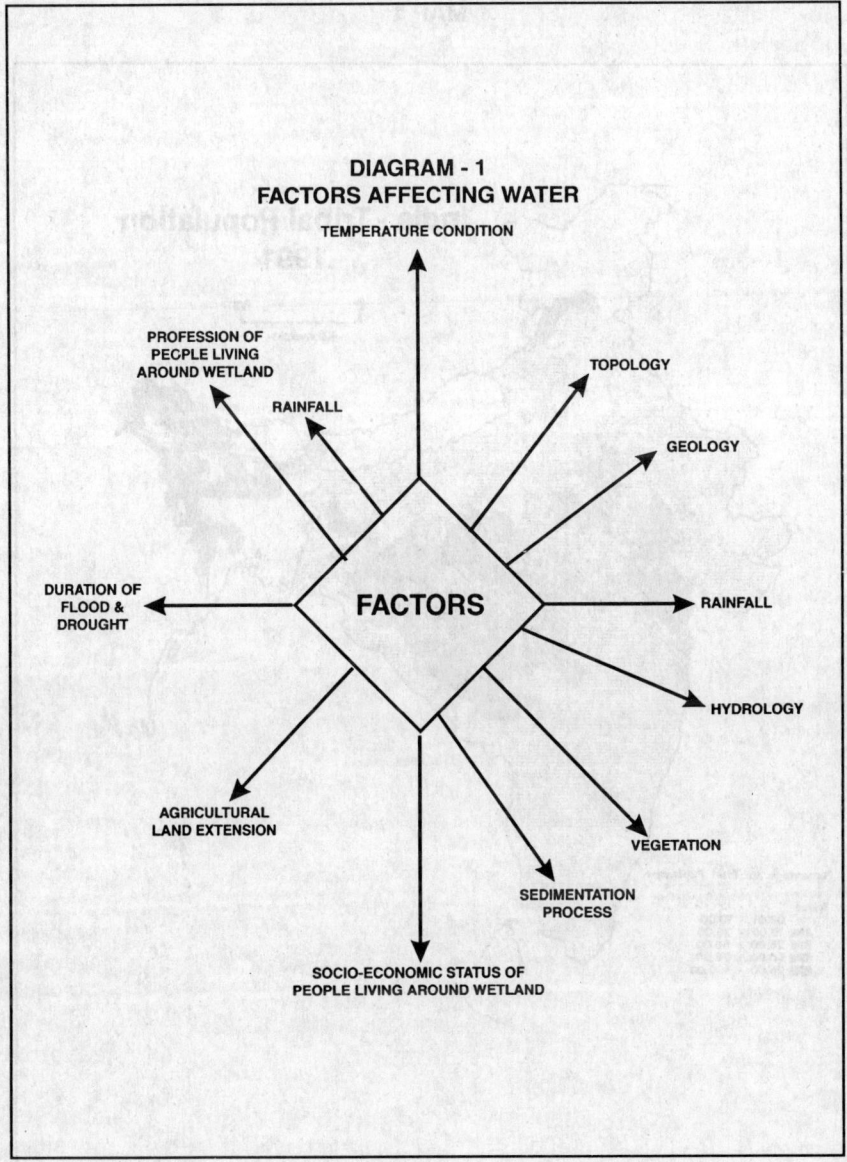

DIAGRAM - 1
FACTORS AFFECTING WATER

MAP 1

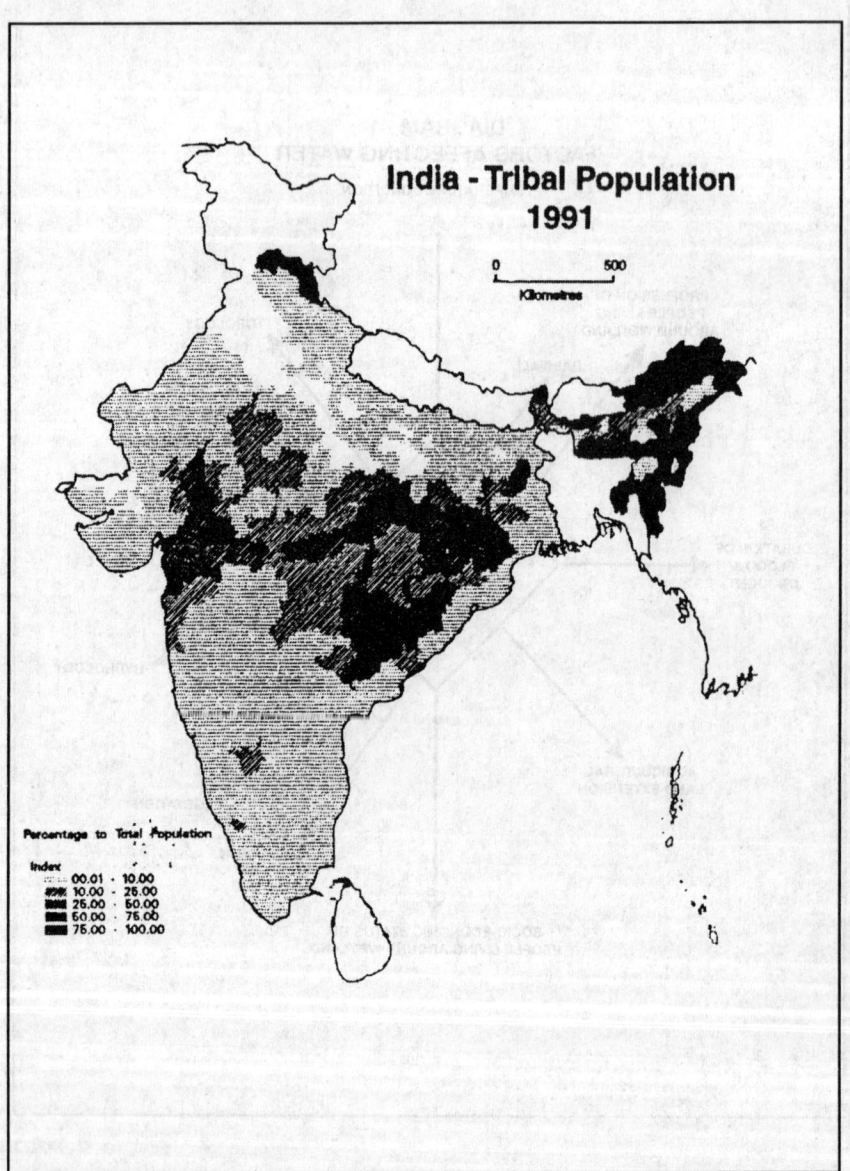

India - Tribal Population
1991

Tribal area, forest cover area mining and industrial areas presented in several maps expose various carcinogenous spots of the country which need immediate steps for greening the areas.

MAP 2

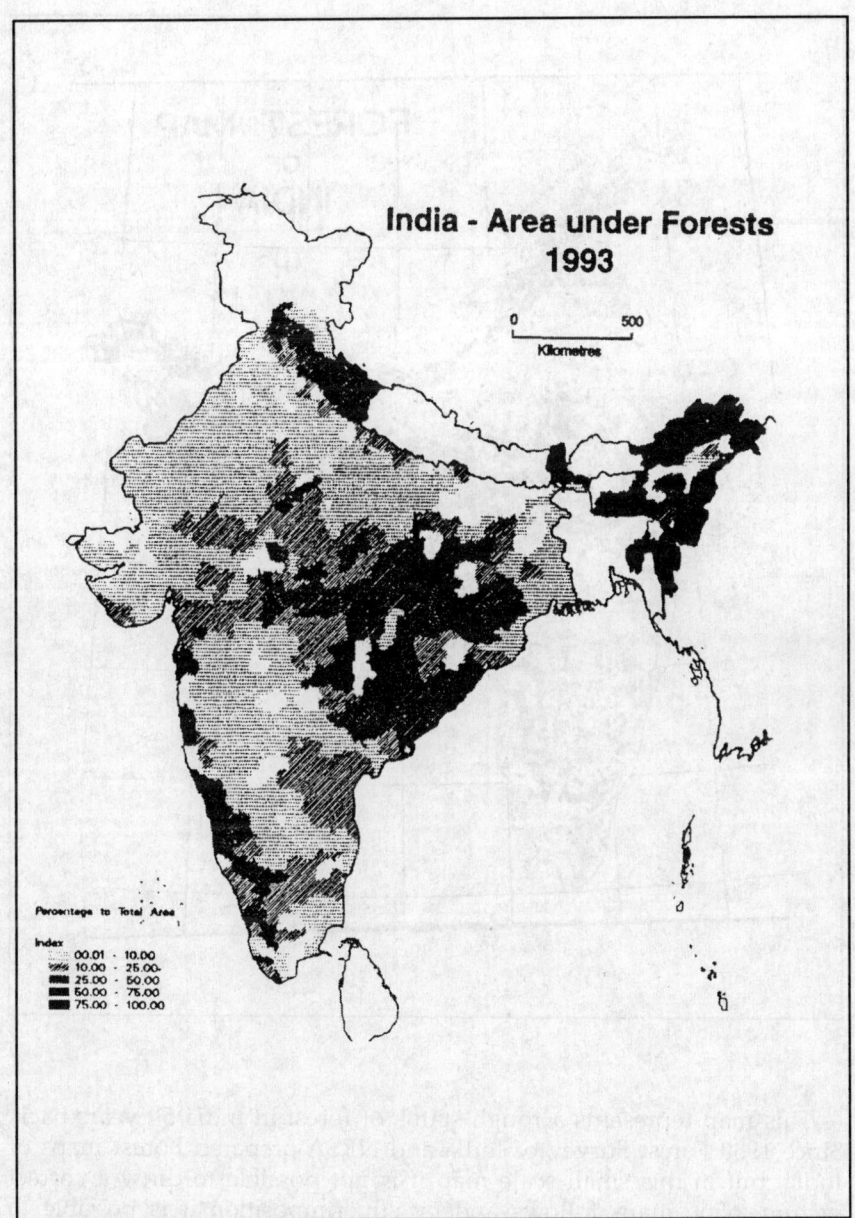

India - Area under Forests
1993

This forest map drawn by another agency presents some what a different picture of forest areas of the country.

MAP 3

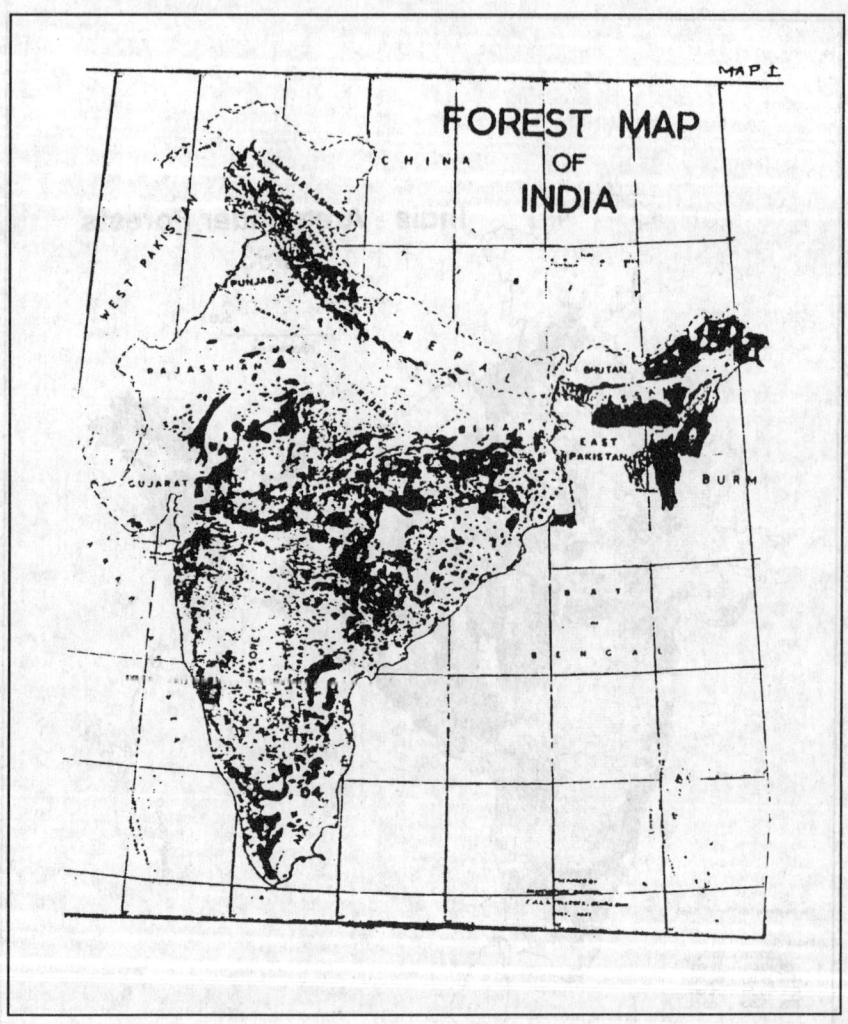

This map represents a rough status of forest in India 50 years back. Since 1980 Forest Survey of India and NRSA prepared Forest maps of India, but in this small scale map it is not possible to draw a correct picture. Nine maps follows and by superimposition it is possible to mark the areas vulnerable to pollution and related problems. Maps of drought prone area, flood affected and soil eroded areas will expose other problems of the country.

MAP 4

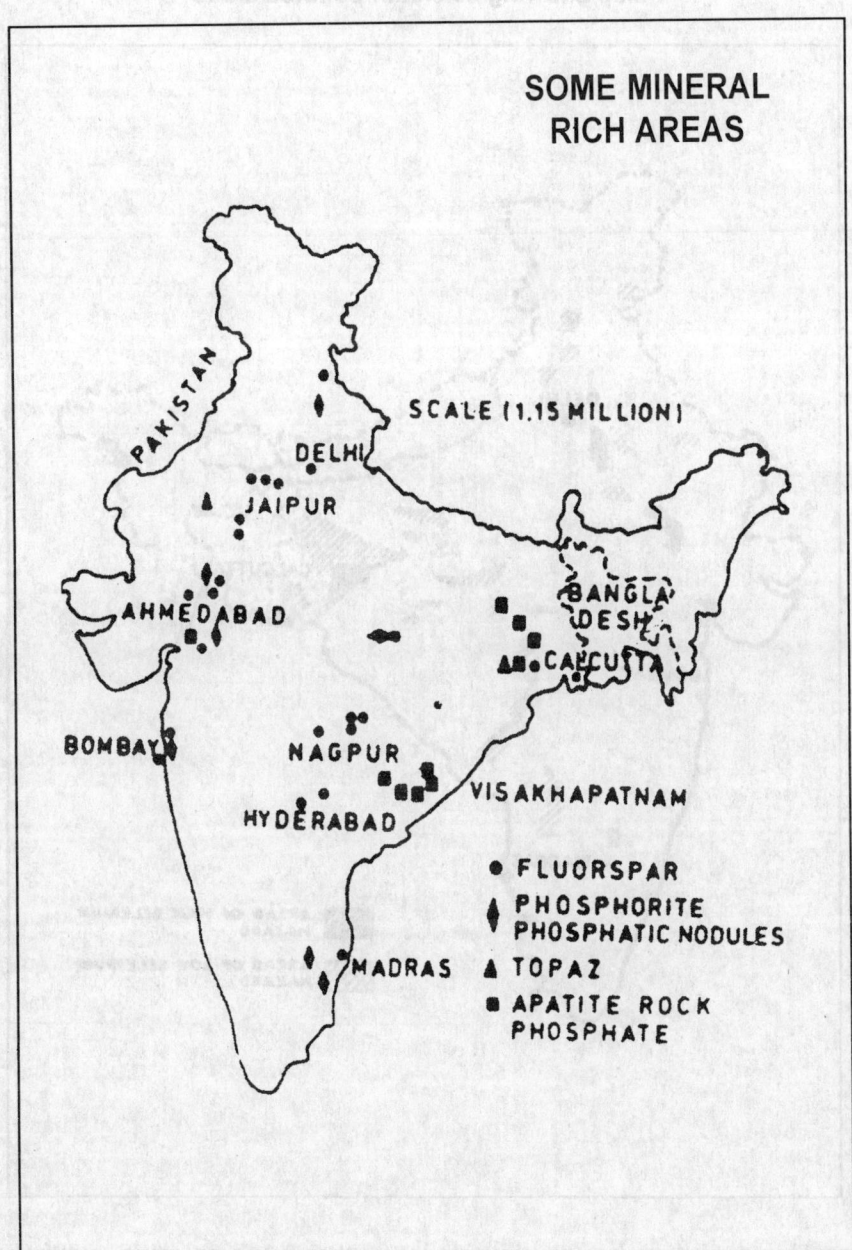

SOME MINERAL RICH AREAS

SCALE (1.15 MILLION)

PAKISTAN

DELHI

JAIPUR

AHMEDABAD

BANGLA DESH

CALCUTTA

BOMBAY

NAGPUR

HYDERABAD

VISAKHAPATNAM

MADRAS

● FLUORSPAR

▲ PHOSPHORITE PHOSPHATIC NODULES

▲ TOPAZ

■ APATITE ROCK PHOSPHATE

This map shows some mineral-rich areas of country.

MAP 5
Map showing selenium polluted areas

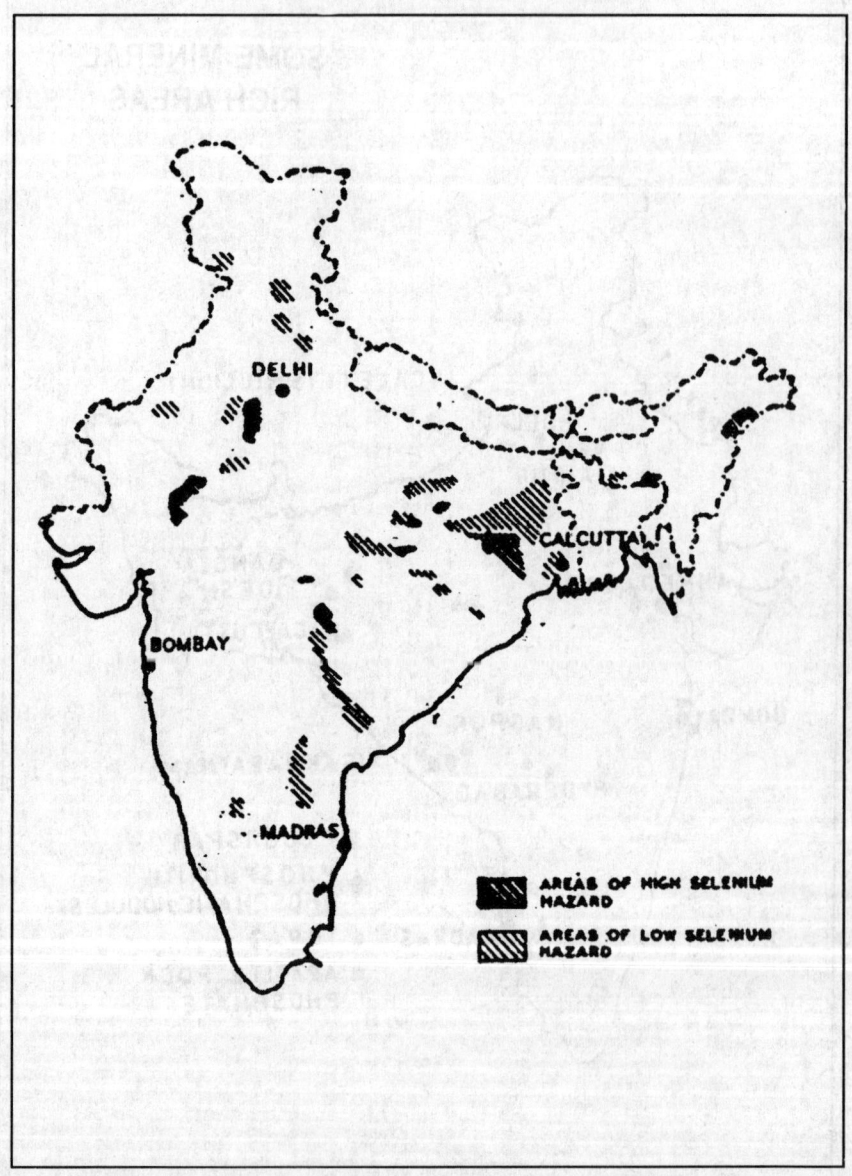

This map may be superimposed on other maps on mine, mineral, thermal power site maps and the depth of the environmental problems assessed.

MAP 6

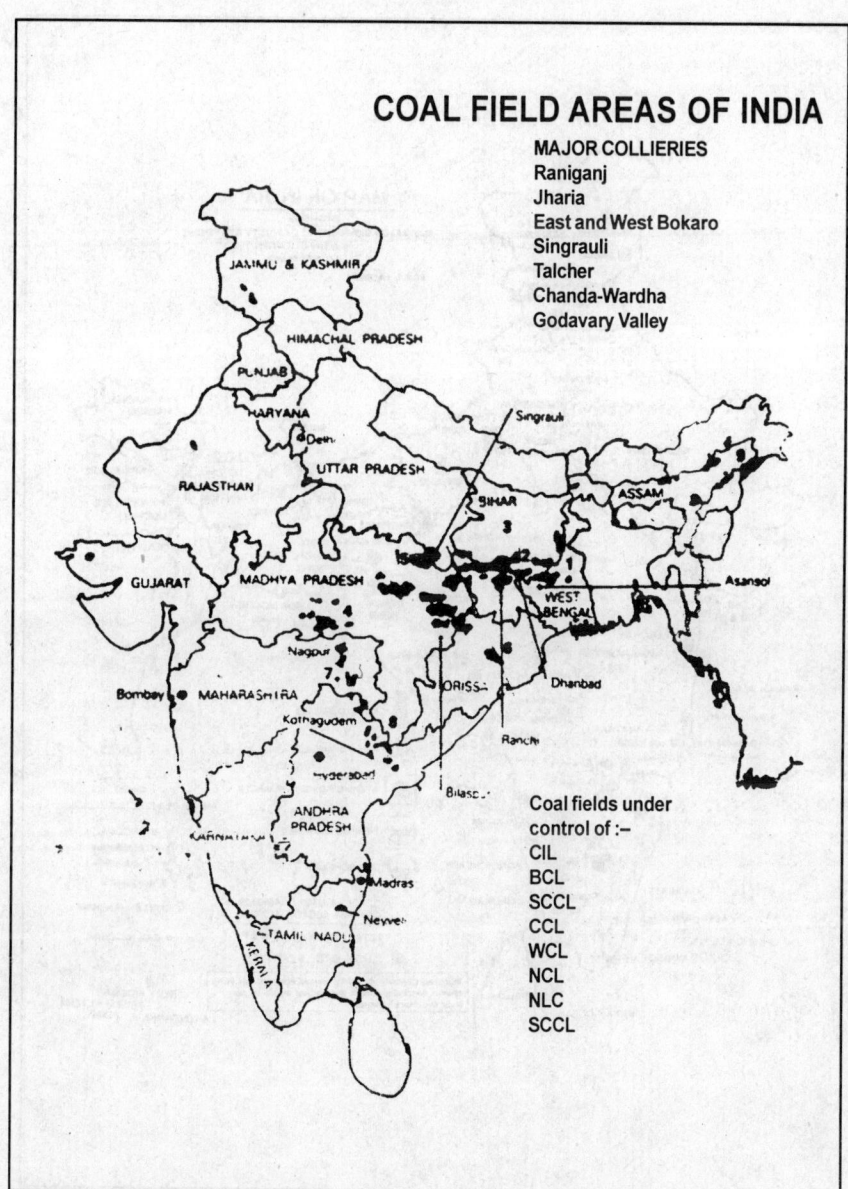

COAL FIELD AREAS OF INDIA

MAJOR COLLIERIES
Raniganj
Jharia
East and West Bokaro
Singrauli
Talcher
Chanda-Wardha
Godavary Valley

Coal fields under
control of :–
CIL
BCL
SCCL
CCL
WCL
NCL
NLC
SCCL

Map showing major coal belts of India. Foresters have a very wide field of works to perform in the form of soil conservation and afforestation to ameliorate the disturbed environment.

MAP 7

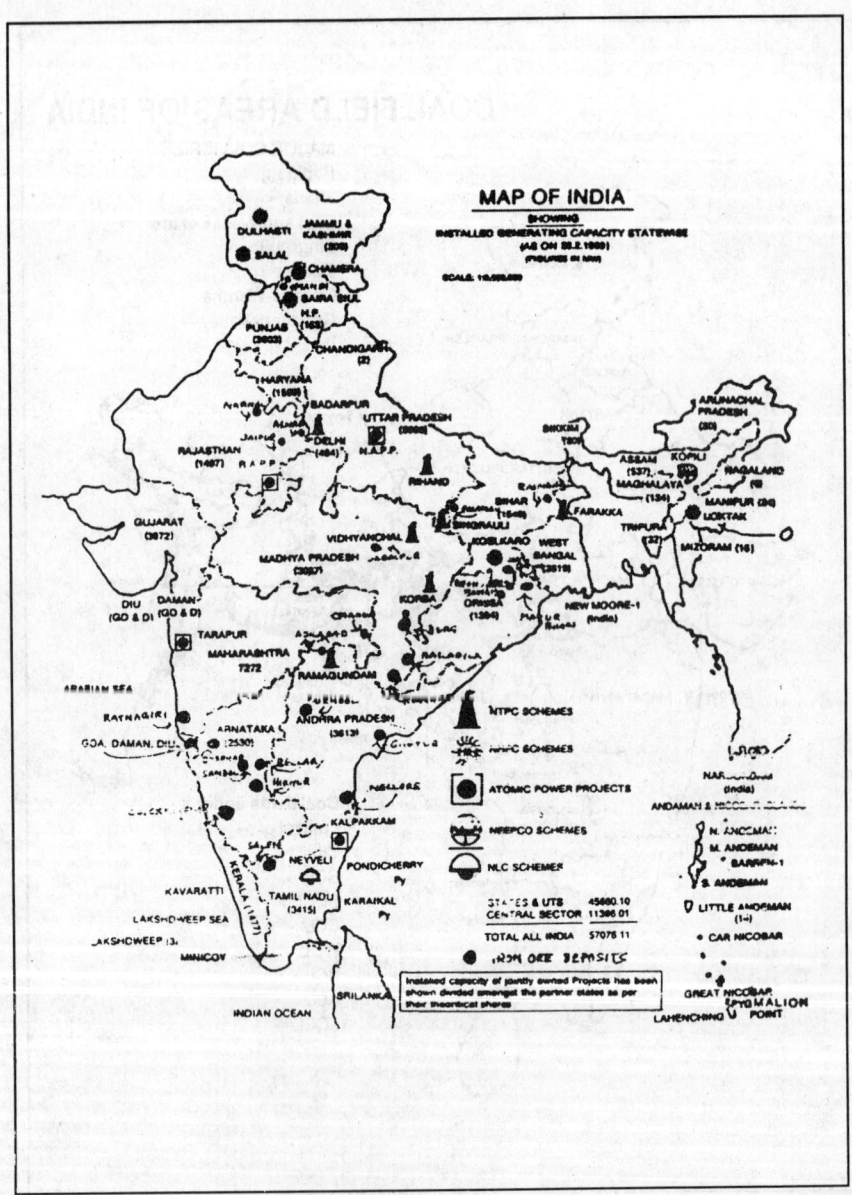

India wide various energy sites are shown on this map. Pollution emanating from these plants are to be ameliorated by vegetal planting as much as possible.

MAP 8

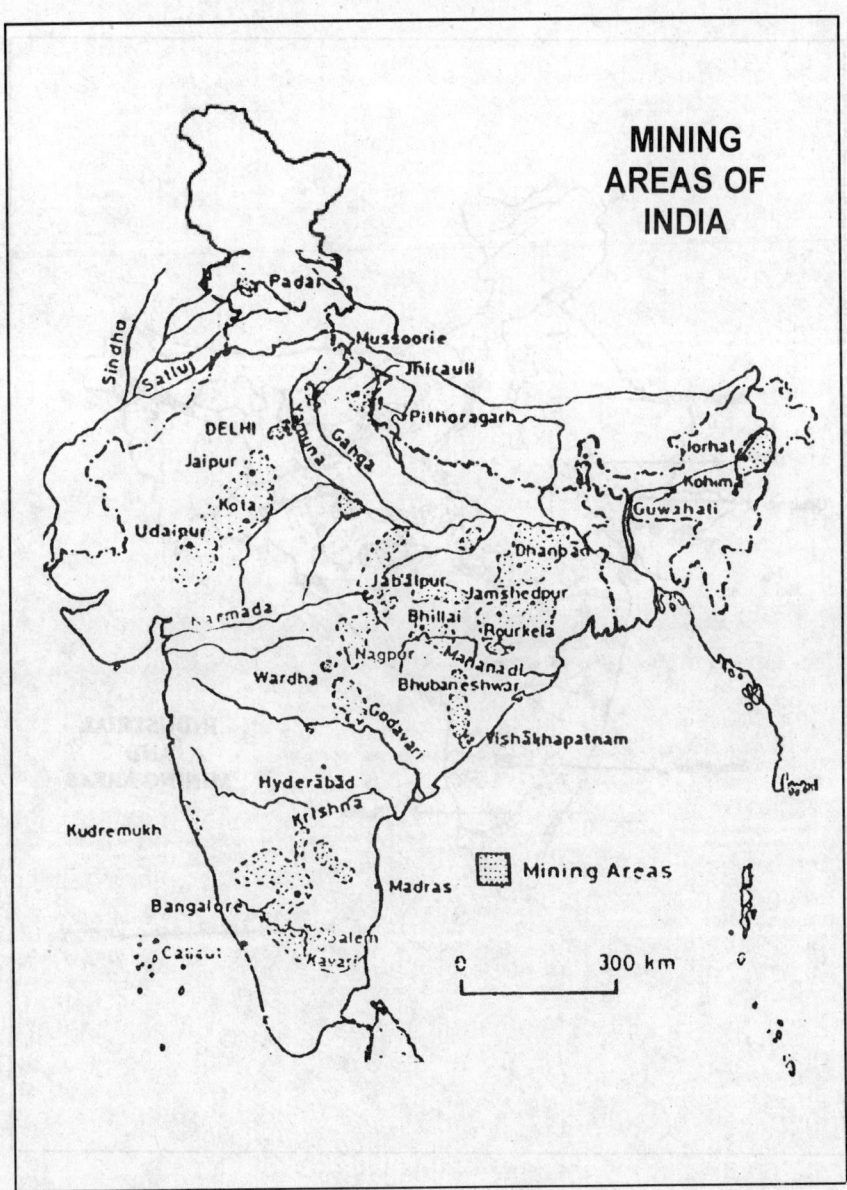

This map broadly shows the mine areas of this country. The map has to be read along with other maps and steps taken to rehabilitate and ameliorate the environment through various forestry methods.

MAP 9

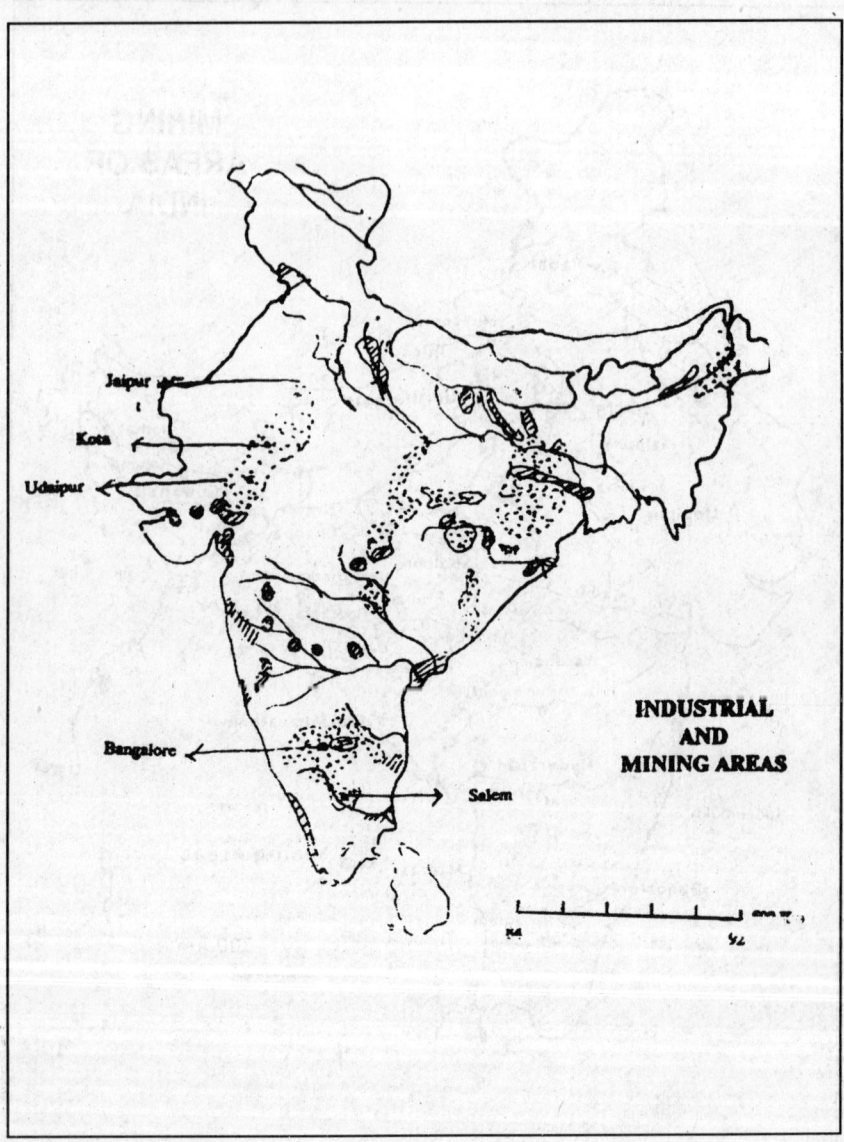

A rough presentation of industrial and mining areas of the country has been made on this map. The cumulative affect of all the issues presented in the maps 4-9 has to be assessed and remedial methods adopted. Foresters have a great role to play in amelioration.

MAP 10
Map showing the drought prone areas

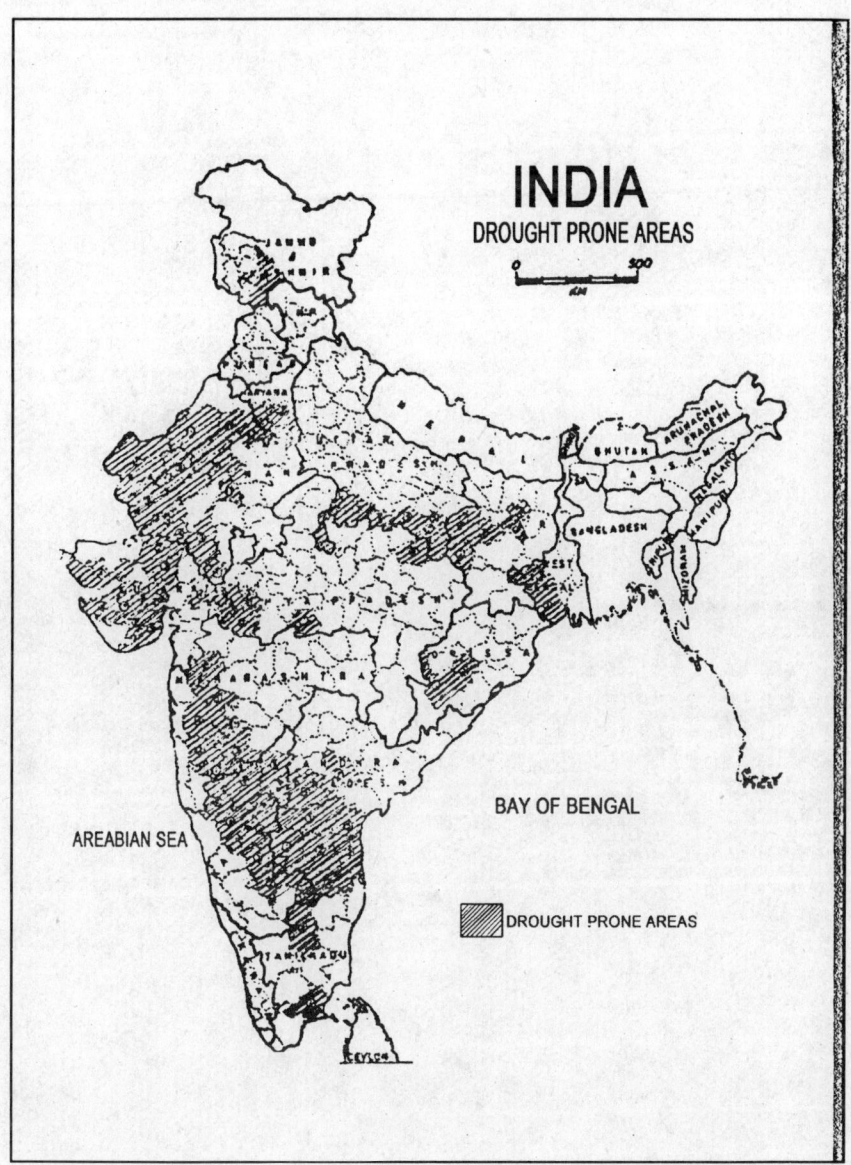

This map represents the drought prone areas of India which very frequently become the victim of drought and sometimes severe drought conditions.

Forest Tribal Areas and Pollution

— Tribals are Exposed to Forestry and Mining Operations and Pollution Therefrom

*T*ribal population are very much dependent on forest and forest resources. Depletion of the forest resources of the country severely affected the life style of the tribal people. In some parts of the country shifting cultivation practices adopted by the tribal people has resulted in depletion of vast forest areas and the ill-effects of such degradation had a tremendous negative impact on the socioeconomic conditions of the tribal population. Moreover the exposure of their society due to mining, implementation of River Valley projects, etc. also invited severe problems in respect of ethnobiology, transmigration, air and water pollution, etc. thus causing tremendous negative impacts on their day-to-day life. Monoculture practices in the forestry deprived the tribal people, living in the vicinity of the forest areas, of getting their age-old requirements of minor forest products.

In this chapter the above issues have been discussed in details and as remedial measures extensive social forestry practices have been suggested. Mass scale afforestation in and around tribal areas is the first precondition of raising the living standards of the tribal people of the country.

Forest Tribal Areas and Pollution

— Tribals are Exposed to Forestry and Mining Operations and Pollution Therefrom

Forester's activities are interwoven intimately with the tribals and their welfare. A perusal of a few maps presented will testify that most of the forests of central, central-west and north east are tribal inhabited.

ENVIRONMENTAL HAZARDS IN TRIBAL AREAS AND THE FORESTER'S INVOLVEMENTS

On being requested by the Birsa Agricultural University, Ranchi to participate at a symposium held in March 1989, the author studied the problems of tribals thoroughly and contributed (also spoke by presentation of slides) a paper entitled "Social Forestry to redefine strategy to assess and ameliorate environmental hazards in tribal areas".

The article is presented hereinafter with slight modification which may be found very relevant to assess a foresters' role of this problem.

There has been monumental work on this subject and the foresters all over the country have created extensive plantations which have improved socioeconomic conditions of the tribal population to a large extent.

Following facets of the issues are enumerated below:

ENVIRONMENTAL CRISIS IN INDIA

The agencies like **"The Brandt Commission"**, **"The Global 2000 Report to the President of USA"** and **"The World Conservation Strategy"** have indicated the environmental crisis in developing countries. The Planning Commission document 'Report Committee on Forests and Tribals of India, 1982' enumerates India's dreaded environmental crisis that the country is reeling under flood, droughts, soil erosion, forest and land depletion, that there is famine in the supply of fuelwood and timber. The report on National Remote Sensing Agency (NRSA) revealed about alarmingly low forest cover (only 14 per cent) and that

1.5 million ha. of forests are being depleted annually. Shifting cultivation in Tribal areas is contributing substantially to this crisis.

TRIBAL DEMOGRAPHIC SCENARIO

Anthropometric information on the tribals of India in the tropical forests is inadequate. A satisfactory assessment of the environment and the socioeconomic and working condition of the population is yet to be made. There is need for extensive studies in physiological characteristics of physical work capacity and heat tolerance, about their migration, adjustment, physical adaptation, method of food intake, nutritional status, etc. for proper assessment of tribal environment.

Many people living in or in the vicinity of forests, in recently deforested areas, may have modified or radically altered their living condition changed with either improvement or worsening hygiene and be exposed to different infections. Such data would be necessary for proper formulation of development programme and environmental assessment. The investigation requires careful planning, clear aims and interdisciplinary approach.

The tribal dwellers have suffered heavy stress due to deforestation and depletion of forests, raising of monoculture crop, reduction of food sources (plant and wildlife). There has been an increasing geographic, social and demographic scale of economic system including the development of large extractive industries and destructive environmental modifications for the benefit of societies geographically or culturally based largely outside the tropical forests. So there has been an abrupt change in population-environment relationship specially in use-system which are sensitive to population density, or hunting and gathering. Increased migration has caused cultural heterogeneity. Owing to opening of mines and industries in the vicinity of tribal areas they have been suddenly exposed to a changed physical, chemical and biotic environmental stresses; sudden alteration of diet might have caused profound disturbance of water and electrolyte balance and caused various diseases. The migration to a new environment might have deleterious effect on social and economic status including poor housing, sanitation, water supply and exposed to many infectious diseases.

There are more than 400 different ethnic communities forming about 38 million population (1971 Census) which increased to 42 million when the area restriction relative to the demographic distribution of the Schedule tribe population within a state was removed. It forms about 6.4 per cent of the population of country. Nearly 82 percent of this population is concentrated in central and western part of the country, about 11 percent in the northwestern states and about 7 percent dispersed in small pockets in the southern zone. Statewise distribution is Madhya Pradesh (20.13 p.c.), Orissa (23.11 p.c.), Assam (10.99 p.c.), Andhra Pradesh (3.8 p.c.),

Bihar (8.35 p.c.), Gujarat (13.99 p.c.), Maharashtra (5.86 p.c.), Rajasthan (12.13 p.c.), Arunachal Pradesh (78.85 p.c.), Mizoram (94.28 p.c.), Manipur (31.13 p.c.), and West Bengal (5.72 p.c.).

The Planning Commission document (1981) considered the hill area of U.P., West Bengal, Assam as backward areas though exclusively backward areas are hill states of Jammu and Kashmir. Himachal Pradesh, Manipur, Nagaland, Mizoram, Meghalaya, Tripura, and Sikkim and also the hill areas above 600 m in the Deccan belt has been considered as backward areas.

Some principal tribes are: Bhils of Gujarat and Rajasthan, Nagas of Nagaland, Gond/Agarias of Madhya Pradesh, Khasis of Meghalaya, Santhals of Bihar and West Bengal, Khariars of Bihar and Orissa, Savaras of Orissa, Lepchas of Sikkim, Todas of Tamil Nadu, Konda Beddis of Andhra Pradesh, Kolams of Adilabad, Nissis of Subansiri and several others.

COMPLEXITY OF STUDY IN TRIBAL AREAS

The complexities are due to, (I) Close correlation between the tribals' cultural attainments and habits with the habitat they live in, (II) their lifestyle and existence depend on favourable physical environment (which also include their house, food and tools, etc.) (III) their evaluation has parallelism with floral and faunal evolution in the same habitat for centuries, (IV) These are from four major stocks of mankind: Negroid, Astraloid, Mongoloid and Caucasoid, (V) They have three major families of languages viz., The Austro-Asiatic, the Dravidian and the Tibeto-Burma, (VI) there are various patterns of societies, villages, marriage types, ethnological groups, nomadic groups, (VII) they have occupational diversity and diverse source of economy, (VIII) They have symbiotic relationship with forests, (IX) They are primitive societies which have achieved some kind of adjustment between its material needs and the potentialities of its environment, (X) They have several sizes of social group, varying material needs and various degrees of skill to tap the resources.

Moreover, plants are intimately connected with rites and rituals. Many of the leaves, fruits, etc. needed in magico-religious rites, are procured from forests. Many indigenous forests are considered abode of deities and are worshipped.

All these make any assessment extremely difficult.

ENVIRONMENTAL IMPACTS ON TRIBALS

Impact of Tribal Renaissance on Sociocultural Scenario

- Several tribal areas have undergone administrative changes. Reorganisation of states took place in several states. Also ecology, economy, social systems, religion, value systems suffered some change.

- Development of road, concrete house, railways, airports, motor cars, cinema houses, schools, colleges, health centres, factories, mineral and non-mineral mines, welfare centres, etc. have affected the root of socioeconomic cultural scenario.
- Vast number of mines were thrown open and expanded. This brought employment, improve communication system and welfare facilities and change in the pattern of living.
- Owing to growth of population, there has been increased requirement of fuel wood, bamboo and timber and this led to ecological depletion and dilution of diversity and density.
- With the awareness about ecology, the clamour for stoppage of Jhum cultivation increased and schemes were started to regulate and replace Jhum cultivation for which land capability and soil survey studies have been undertaken. Barren areas were brought under coffee, rubber, tea, pineapple and various other horticultural crops. All these activities are fast changing the life style of the tribal people. This change is faster in north eastern states owing to more spread of education and more awareness of a better future.
- Various schemes have been introduced by the Govt.. to improve this socioeconomic conditions of the tribals and therefore society structures are also fast changing.
- The sixth schedule allowed the formation of District Council run by educated persons. More education and more civic facilities were extended. During the second and third Plans the development of villages and small scale industries became the land mark of industrial development.

These rapid changes might have brought about undesirable cultural disorganisation and some sort of resistance against industrialisation which need a thorough assessment.

Impact of Transmigration

India is intent on developing the country's isolated and alien groups. To achieve that end the tribals sometimes are resettled under transmigration programme. This is to raise their living standard and incorporate them into the mainstream of Indian society. But this may mean their alienation, poverty and cultural death. They are being turn out of the social and cultural fabric that has given their lives meaning for thousands of years.

- In funding various transmigration schemes the World Bank is breach of its own economic and environmental policies in some countries.

Some environmentalists linked this issue in Indonesia as an environmental impact to "Waging the equivalent of thermo nuclear War" and "as the Single Sectorial activity with the greatest potential to advance forest destruction-often to no constructive result".

- The Dandakaranya Project was human and natural disaster where the settlers never settled properly and the tropical forests of Baster, Koraput and Dhenkanal districts lost species diversity, exposed sterile soil caused species extinction and nutrient deficiency of soil. A vast area of 75,000 sq. km of tribal reserve where the Gonds, Khonds, Jalapas, Bhuiyas, Battadas tribes used to live was destroyed.

- These tribes must have faced a large scale impact on their family, education, caste, attitude, health, religion, consumption pattern, food and migration pattern.

- In erstwhile Bhopal State rehabilitation of Bhils, Gonds, Korki, Keer, Mongra, Sansia tribes were made. Many landless were settled. Similar rehabilitation was done in the states of Bastar, Sarguja and Jashpur.
  B. Chowdhury in his book "Tribal Development in India" mentions about displacement of tribals (28,275 families) from 216 project areas during Second and Third Five Year Plan involving 4210 ha. of land.

- The Bhillai Steel Plant, the Korba Thermal station, the Parasia Coal Mine, Kolma Collieries are located in tribal areas and the tribals had to migrate elsewhere.

- Several Chinchu tribals are to be shifted from Amrawati Tiger Reserve.

Impact on Ethnobiology

Ethnobotany is defined as "Systematic study of the botanical knowledge of a social group, its use of locally available plants as food, medicine, clothing and religious rituals". It is also the study of cultural use of plants. Ethnozoology has the similar definition. The studies made by Botanical Survey and Zoological Survey of India have recorded widescale depletion of such resources. Several Research Projects have been initiated by the "Ministry of Environment and Forest"; such studies have revealed that a veritable arsenal of wild plants with potential for exploitation for humanity is available in the forests of which the tribals/aboriginals have specific knowledge. Over 1900 plants and animals had been recorded in 1984 used by the tribals for food, fodder, fibre, medicine, clothing, shelter, etc. The Z.S.I. has recorded mammals and birds used as medicine by the tribals; they use blood, bone marrow, liver, urine, excreta as medicine.

The conservation of plants and animals of the forests had been a part of their faith and traditions and a set of fables, folklores, totoms, taboos, religious belief perpetrated this culture.

Impact of River Valley Projects

- In 1987 the Govt. of India granted the states of Madhya Pradesh and Gujarat permission to transform the Narmada River and it's valley through a series of 3000 associated dams, 135 minor and 30 major dams. The principal focus of attention was the two giant reservoirs, the Sardar Sarovar in Gujarat and the Narmada Sagar in Madhya Pradesh. About 3,50,000 ha. of forest land and 2,00,000 ha. of cultivated land would face irreversible environmental changes. The tribal residing in the area are the Bhils, and Gonds and the Banigas. About 10 lakh people would likely to be affected of which substantial population would be tribals who have to be transmigrated.
- The Bodhghat Hydel Project in Bastar would affect about 9000 tribals who have to be transmigrated from the submerged area (13,526 ha.).
- "Hydel Project" or "Cultural Ethnocide".
  The Munda tribals are fighting against Koel-Karo project in Chotanagpur plateau. Munnar in Kerala, Bedhhi in Karnataka, Tehri and Vishnu Prayag in U.P., Lalpur in Gujarat are other projects but fewer tribals will be disturbed. Propatpatnam and Inchampalli in M.P., Maharashtra, Andhra Pradesh border will drown about 2,00,000 ha. of land and effect 40,000 media Gond tribes who would have to be transmigrated to Marathi speaking area. Baba Amte considers this as *"Cultural Ethnocide"*.
- About Dihang the Chief Minister of Arunachal Pradesh remarked.

"We cannot agree to a project like Dihang as it would affect the life style, culture and environment of almost one sixth of the population of this sensitive area".

Impact of Shifting Cultivation

Over more than 1,00,000 sq. km. of forested land in the states of Orissa, Madhya Pradesh, Andhra Pradesh, Bihar and all the states of north eastern India have suffered from deforestation and land degradation, and depletion of water regime. With the pressure of population Jhum rotation has become shorter and shorter and has led to soil deterioration, loss of species diversity, depletion of wild life and eventually loss of ecological balance. According to Planning Commission (1982) 233 blocks spread over 62 districts in 16 states the Jhum is practised involving 12 p.c. of tribal population.

Names of the tribes in various states involved in Jhum area:

Andhra Pradesh	–	Bagata, Gandabas, Khonda, Koyas, Konda, Beddis and others.
Assam	–	Mikir, Mizo, Naga, Chakma, Jainlia.
Bihar	–	Malpaharia, Birjea, Korwa.
Gujarat	–	Bhil, Kunbi, Kokanis.
Madhya Pradesh	–	Agariya, Baiga, Bharia, Gond, Kowra, Media, etc.
Maharashtra	–	Marja Gond, Thakur, Kunbi, etc.
Orissa	–	Bhuiya, Gadabas, Kond, Koya Kotia, Saor, Saora, Paroja.

Districts other than north-eastern India where Jhum cultivation is done are:

Andhra Pradesh (East Godavari, Srikakulam, Vishakhapatnam) Bihar (Santhal Parganas), Madhya Pradesh (Baster), Orissa (Dhenkanal, Ganjam, Kalahandi, Keonjhar, Koraput, Sambalpur and Sundergarh).

Impact of Mineral Environment on the Tribals

Although mines of India occupy an insignificantly small area the impact of mines and mineral based industries in tribal belt is significant and it is an imperative obligation of the administrators to restore, rehabilitate and revegetate such areas by biological and engineering processes.

A superimposition of three maps, viz. forests, tribal inhabitants and mineral resources will testify to the fact that mineral and tribal areas are interwoven. As such, this insignificantly small area (relating to the geographical area of India) inhabited by various tribal communities form a carcinogenous nucleus in a tribal environment.

Some observations are as follows :

Subarnarekha Basin

The basin, the smallest of fourteen river basins of India, is a typical example of victim of hazards brought about by mining of a good section of 11 lakhs schedule tribe and 9 lakhs schedule caste population.

Once a quiet coy with crystal-clear stream blooming with lotuses hide behind primeval forests, the Subarnarekha is now rocking under booms of dynamite charges and the rattle of heavy earth moving machines, continuously unearthing its rich mineral wealth and demolishing the ramaining forest cover and the soil mantle all over the basin eroded.

- Strip mining, mine tailings, ore dumps of copper mines, radio active waste materials in the uranium mines at Jaduguda contaminate ground and surface water in tribal areas.

- Processing of Uranium ore which are converted into magnesium diuranate by wet process contaminates more areas.
- The Apatite (which contain fluorine) ore causes fluorine contamination in surface and ground water system.
- At Rukha (near Ranchi) Subarnarekha contains 0.74 milligram of fluoride per litre of water.
- Gorumahishni – Badampahar ranges have a large number of abandoned mines, unprotected high benches on the hill slopes are polluting both land and water at lower level.
- Carrying of Chinaclay, Limestone, etc. also cause wide scale pollution.
- Analysis made under the auspices of Global Environmental Monitoring System (GEMBS) at Ghatsila and Talisilwai in Subarnarekha show low (2 mg per litre) dissolved oxygen content.
- Copper deposits have other associate elements such as Sulphur, Nickel, Cobalt, Molybdenum, Selenium gold, Titanium, Silver, Platinum and Uranium. Rain water percolating through the open sulphide ore dump and huge dumps of rock wastes leach toxic elements like, Selenium, Molybdenum and Uranium. Some toxic elements find way to agricultural land.
- Vanadium at Dublabera, Kumbhar Dubi, Bisai (Mayurbhanj District) Chromite at Ruruburu, Chilungburu, Kimsiburu and Jojohatu, China Clay between Kharkai and Vaitarani basin put substantial land out of use. Tailing of asbestos mine does not bear any vegetation.
- Chromium as Chromite has minimal toxicity, but should it be oxidised to hexavalent Chromium the tailing could become very toxic and may be carcinogenic.
- Iron and Manganese oxide play an important role in the soil in fixing trace elements such as Cobalt, Copper, Zinc, Nickel as well as pollutant like Lead. The associates of these elements with Iron and Manganese has important implications for agriculture and plant growth. Once fixed, the elements are unavailable to plants. High pH, excess of Phosphate Bicarbonate, Copper, Zinc, Cobalt, Cadmium, Manganese or Nickel in the growth medium may cause iron deficiency.

So the tribals residing in the Subarnarekha basin have to use contaminated water, their agricultural land has been rendered unfit for use and workers face noise and air pollution.

Tribals in Orissa Mines

- Orissa has 237 mines (Keonjhar 66, Sundargarh 81, Cuttack 16, Sambalpur 38, Mayurbhanj 18, others 18) and about 80,000 people are employed there. An investigation in 28 mines in Keonjhar and

Sundergarh districts showed about the tribal engaged is about 85 per cent.
- Safety rules are not observed meticulously and the mines are exposed to various types of hazards such as noise, dust and polluted water.
- Besides, a good percentage of Mayurbhanj's 58 percent, Sundargarh's 53 percent, Keonjhar's 46 percent, Koraput's 56 percent of Tribals are exposed to hazards of mining.

Industrial pollution; it will be a big list for mentioning of all the items as it will become more voluminous.

Some of the industries are:

Ferro-Manganese Plants at Rayagoda and Joda, Pig Iron Plants at Joda, Ferro-Silicon and Silicon Metal at Theruvali, Cement Plants at Bargarh, Fertilizer plants at Rourkela, Thermal Power plant at Telcher, Sponge Iron plant near Palaspanja, etc.

Toxic Selenium hazard

Selenium is present in sulphide material, coal, phosphorites and uranium bearing rocks and soil derived there from. Selenium is responsible for certain acute and chronic diseases in man and animal. Selenium rich zones are Singhbhum, Aravalli group of rocks, Meghalaya, Upper Assam, Damodar Valley, Jabbalpur, Mahanaeti Valley, Satpura hills and Guddapa basin.

Tribals and Mining in Central India

The tribals in Central India, whose lands and forests have been extensively destroyed by mining and associated industry are worst sufferer. No less than eight super Thermal Power Stations have been planned in a cluster to the coal mines of Singrauli. Similarly all major steel plants are concentrated in Bhilai. Bokaro-Rourkela-Jamshedpur, Durgapur, Burnpur region. Chattisgarh was once a rich bowl with fertile land now poverty stricken people move about under oppressive contractual condition.

Mica Mines in Bihar

About 2,50,000 tribals are engaged in Mica mines who brave various hazards of mining.

Cement Factories

About 20,000 workers at Khinkpani, Chaibasa belong to Hoand Diker tribes mostly.

- Chittaranjan locomotive works has Santhals and Kol tribes, residing in the vicinity of the workshop.
- Narmada Valley has rich coal, iron, limestone, copper, dolomite, manganese, Bauxite and soapstone materials extended over tribal areas.
- A lot of tribal workers are engaged as unskilled workers in various industries at Amar Kantak, Bharat Aluminium Company, Hindustan Aluminium Company, etc.
- In Balaghat district Malanjkhand copper mine point is causing serious ecological problem. 10,000 sq.km. of coal area and 14,000 sq.km. of Bauxite area cause various hazards to the tribals, residing in the vicinity.
- Extraction of lead, zinc, tungsten, gypsum, asbestos, etc. in Rajasthan is responsible for soil salinity of extensive area.

Impact of Air Pollution on Tribal Areas

Several instances have been cited above about the hazards of various mining operations faced by the tribals. Wind carries dust from waste heaps, blasting and heavy machinery operated carries dust over extensive area. Blasting produces obnoxious fumes. These produce respiratory diseases and eye ailments. Pneumoconiosis, silicosis and asbestosis are caused by inhalation of dust. Common disease that follow are cataract, conjunctivitis, corneal ulcer, glaucoma and squint trachoma.

About Roro Asbestos mine at Chaibasa I.L.O in 1978 remarked "the Ror mine endangered not only the lives of mine workers but those living in the surrounding areas". Ore concentrators, crushers and smelters are further sources of pollution.

Impact of Land

Agricultural land degradation has been mentioned in above. Limestone, Quarries and Cement factories have reduced land productivity in areas of Udaisagar, Khamli and Chittor in Rajasthan, Agricultural fields in a 10 mile radius around the cement factory at Khikpani in Singhbhum lie barren under layers of accumulated dust.

Impact of Water Pollution

Ample instances have already been cited in foregoing paragraphs on water pollution. Ore fines and toxic substances carried by rain water into water courses alters their chemistry and make it unfit for human consumption. The untreated effluents, slime or tailing are released in neighbouring stream.

- Effluents from Bailadilla Iron Ore mines containing 18 percent of mined ore are dumped into nullah. The turbid streams join the

Sankini river turning its water red. This water is used by a population of 41 villages with 40,000 villagers mostly tribals, who have named the river as "The River of Sorrow".

- Continuous dewatering under ground of mines (millions of litres) cause flooding, silting, water-logging and pollution. Water table is lowered.
- Underground waters of Birsinghpur, M.P. are pumped out 3000 litres of water per minute into Ganyra nullah. The villages of Neyveli lignite mines have protested against such dewatering.
- The Chotanagpur plateau and the Maikal Range feed five major rivers – the Narmada, Sone, Rihand, Mahanadi and Damodar and the area is vulnerable.
- At Bastar the National Mineral Development Corporation (MMDC) may get lease of 800 ha of Kanger Reserve for Dolomite. This may cause concern to local tribals.
- The relation between agriculture and trace metals is to be solved as elements like cadmium or lead which does have biological function is potentially harmful.

SOCIAL FORESTRY

The senior author's assessment of the subject is up to 1989. By this time there has been a lot of achievements in this sphere since 1989. However, it started with an aim:

- To meet the most urgent requirement of fuel wood in shortest possible time and to provide poles, timber, bamboo, fodder, fruits and other minor forests products for basic requirement of rural population.
- To provide employment to the unemployed and under employed local people, particularly the landless agricultural labours including tribals, schedule cast and other traditionally weaker section of the rural communities.
- To induce community participation in creating, maintaining and protecting plantation programme under a joint management system with share of benefits of the produce raised.
- To improve the environmental condition by creating green belts in urban and industrial areas thereby ensuring aesthetic and recreational facilities.
- To supply improved variety of seeds, grafts, seedlings for increased production and up-liftment of the economic standard of local people.

Various objectives have met with commendable success in West Bengal. Similar success has been achieved in Gujarat, Uttar Pradesh, Tamil Nadu and some other states. But the aspect of improvement of environmental condition in creating green belt in urban and industrial areas leaves much to be desired. Some tangible benefits have occurred due to planting on non-forest land where soil and water conservation status are improving the quality of life.

Lately, there has been a new strategy to allow the participant people with part or full share of the yield. Generally the following components are built in any social forestry programme:

- Reforestation of degraded Government Forest Land.
- Community forestry in barren land.
- Farm Forestry on privately owned waste land.

This has created an atmosphere when more and more people are being motivated and involved in the Social Forestry Programme.

Social Forestry in Tribal Areas

- Social Forestry has so far not been a promising tool for socioeconomic and ecological development, continuous employment, better nutrition, higher quality of clothing, shelter, better health and recreational facilities in tribal areas. Bulk tribal population has not been covered under direct and immediate microlevel benefits to resources – poor and economically disadvantaged rural dwellers, as well as bring forth long term, macro-level resources-conservation type advantages that Government agencies try to achieve. Neither, it has proved useful, as a small-scale landuse operation ranging from pure forests to integrated agro-forestry in tribal areas; nor it has initiated interaction between the forest ecosystem and human social systems in tribal inhabited area. It has yet to deal with broad aspect of human ecology in tribal areas.
- Social Forestry and tribal areas will need to consider social factors such as cultural knowledge and values regarding forestry, availability of resources, land, capital, materials and labour, social constraints on resource management, social competition and conflict over resources.
- Social Forestry, is considered a major element in India's overall record development but it has to cover bulk rural areas of central India.

Why Plant Tribal Inhabited Mine Areas?

The problems of dust, mine wastes, acidity, toxicity, presence of trace elements and heavy metals, carcinogenic materials and other pollution hazards in mine areas over about 6000 mines have been identified. The role of vegetation to mitigate such hazards has been considered an imperative tool considering its comparatively low investment and lasting benefit. Some benefits of trees planting are listed as follows:

- Dense vegetation intercepts incoming and outgoing radiation and dusty wind, atmospheric particles, create microclimate and ameliorate weather besides supplying fuel wood, fodder, timber, fruit and M.F.P. Such screening is absolutely necessary specially in mining areas.

- One ha. of spruce collects about 32 tonnes of dust from atmosphere, a Beech tree collects 68 tonnes. So creation of barriers of dense tree crop supplemented by shrubs and herbs is essential. Such planting will purify air in mining area.

- Dense ground vegetation (grass, agave, herbs, shrubs) arrests flow of water and help deposition of silt and reduce erosion and water pollution hazard. Leaves absorb sunlight and reduce ultraviolet radiation.

- The vegetation will screen field and pastures against radioactive dust to save people from the effects of radioactivity contaminated nutrition.

- Dense ground vegetation (grass, agave, herbs, shrubs) absorbs pollutants, which are so harmful to health. The sound absorbing quality of the trees will depend on its structure (depth, thickness and types of leaves, etc.). The denser the undergrowth and the ground vegetation and canopied nature of forests, the greater its vertical closure, and higher in its sound absorbing capacity. A dense vegetation of 50 m width or varying height reduces the traffic noise by 20–30 decibels.

- A. Bernatsky's work on the role of protective forests in industrial site, may be considered for reducing hazards surroundings of industrial areas where the tribals live.

- Plants with a high metal or chemical tolerance can be effectively used as indicator and pollution scavenger which have developed in built system for withstanding pollution hazards.

- Plants are selective absorbers of metal ions. The mechanism of toxicity is not fully understood. Some concentrate metals in their cell walls are without any ill effect. In most cases metals disrupt active metabolic sites within the cells.

- Plants have ability to absorb P.A.H. (polycyclic aromatic compound) from air water and soil by roots and foliage. Whereas ordinary phenol is not found in plants in a free state, a wide range of phenol compounds, including catechol, resorcinol and hydroquinone are generally present. Phenolic compound in the protective covers of plants play a role in disease resistance.
Vascular plants have ability to remove phenolic compounds from polluted water. The uptake of heavy metals by marine plankton is significant.

- Although most plants die in the presence of 0.1 percent of copper in soil, it is remarkable that certain plants can tolerate such conditions. Some grasses have high degree of tolerance. This tolerance is confined to only one metal by a species.

- Micro-organism (algae, fungi, bacteria and protozoa) are involved in many key ecological processes such as primary production of organic food materials, and in the cycling of elements. Some microbes are more resistant to metal pollution than higher plants and animals and could, therefore, play an important role on the recovery of polluted area.

Rehabilitation, Restoration and Revegetation in Mine Areas is a Challenging Job

- Mine lands are a fascinating challenge to the foresters, because the preexisting ecosystems are extinguished, it's also a challenge to the biologists and engineers to replace them as they were.
- No adverse environmental impact should last longer than half a childhood, i.e. 5 years.
- The restoration of landscape is generally accepted as an integral part of mining operation.
- We have to develop the means of economically establishing permanent, virtually maintenance free, vegetation cover or submarginal waste, Ore-dump and tailing.
- To achieve permanent revegetation of toxic, sterile and infertile site we have to synthesise an Ecosystem.
- Revegetation entails the establishment, not only of plant species, but of biological processes, such as nitrogen fixation and mineral turnover by decomposer organism, site evaluation, etc.
- In gully control, a bag of fertilizer is more effective than a bag of cement.
- In planting choice of species the help of geo-botanical indicator (pollution scavenger) may be necessary. Adverse effect of toxic metals are reflected in the morphology of plants which may be a guide in assessing status of soil and plants and their interactions.
- Lichens have been recognised as an effective pollution monitors and their analysis may indicate the status of pollution in atmosphere.
- Metal toxicity and their effect on various plant parts have been established and such data may help in assessing the environmental status.
- Every mining of every mineral causes specific environmental problems.

The chemical aspects of soil is a complex study, but such study should be considered essential for any useful planning of revegetation of mining areas.

Initial work of back filling of void, regarding the overburden, proper utilisation of top soil, etc. would be necessary. Besides several other technical works, some other means should be considered necessary before undertaking revegetation work which are – diversion of surface water from the mines and tailing dam areas, drainage stabilisation, collection of all acid affluent to a single treatment either by oxidation, neutralisation and precipitation, disposal of precipitated sludge and the diversion of treated water into the natural stream below the mine working, or by impounding and natural evaporation.

About plants, it has been found that some plants have ability to flourish in elevated levels of trace elements (cryptogams have a high level of tolerance). The tolerance of plant to pollutants can to a certain extent be related to their genetic variability. Some plants have developed inbuilt systems for withstanding pollution hazards.

All these reveal the complexity of the subject.

Social Forestry – to Formulate a New Strategy

Various environmental, social and economic problems and hazards suffered by the tribals have been enumerated in broad outlines in the foregoing paragraphs. It is obvious, the most serious of the hazards are faced by those tribals residing in the central and western India.

The authors consider that in the present economic situation of the country only the forestry sector can bring about a miraculous and most effective changes in the habitat and ameliorate the hazards of day to day life of the tribal communities working in more than 500 mines in the country.

But the job will be extremely challenging as the tribal are sensitive to any drastic change in the habitat, the soil is sterile and the task of revegetation is fraught with various problems. In selecting species various factors enumerated about soil-plant interactions have to be considered.

Social forestry will have to devise planting techniques and choose proper species to check erosion in quarry site, overburdened dumps, spoil dumps, abandoned quarries, tailing dams and revegetate all such areas. For improving quality of life there will be need for planting flowering trees, fruit trees, create horticultural garden, mini zoo, demarcate nature reserve and plant various species to supply minor forest produce.

Suitable species have to be selected to buildup sound and dust barriers. Proper use of scientific methodology in creating shelter belts in all vulnerable areas has to be made. Various low cost soil conservation engineering designs in controlling gully and sheet erosion, controlling nullah, river banks have to be made. Intensive and extensive planting of grass, legumes and shrubs will be necessary.

A very well knitted extension service facilities have to be organised to contact various groups of tribal people to educate them about the hazards they are suffering from and how far tree planting can improve the situation.

Present System of Revegetation in Mining Areas

The present system of revegetation work is done by the mining authorities themselves. They get management plan prepared by experts and competent authorities approve of the plan.

But the authors consider this work extremely difficult. Afforestation work cannot be successfully done by the mining authority and only the experts in Forest Department, well-supported by research base can tackle such serious problems and Social Forestry wing is the best organisation to tackle the situation of vegetation and motivate the tribals for success of the programme.

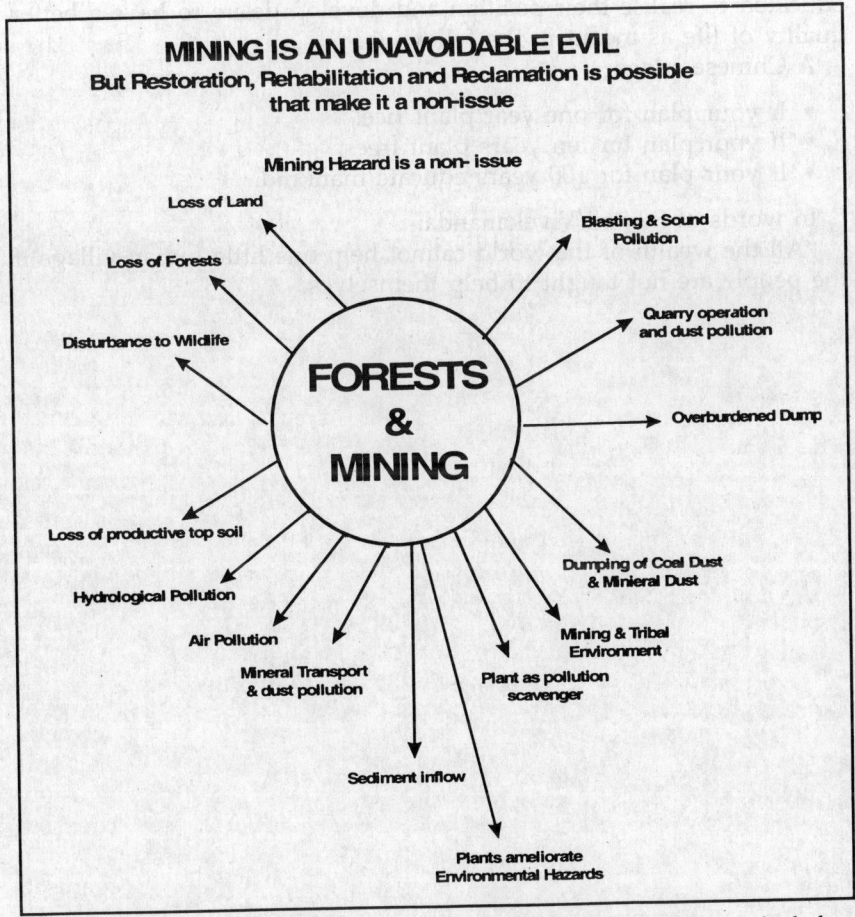

MINING IS AN UNAVOIDABLE EVIL
But Restoration, Rehabilitation and Reclamation is possible that make it a non-issue

Mining Hazard is a non-issue

- Loss of Land
- Loss of Forests
- Disturbance to Wildlife
- Loss of productive top soil
- Hydrological Pollution
- Air Pollution
- Mineral Transport & dust pollution
- Sediment inflow
- Plants ameliorate Environmental Hazards
- Plant as pollution scavenger
- Mining & Tribal Environment
- Dumping of Coal Dust & Mineral Dust
- Overburdened Dump
- Quarry operation and dust pollution
- Blasting & Sound Pollution

FORESTS & MINING

**The mitigation of these hazards is a challenging job of the foresters, which they can ably comply, should adequate fund be placed at their disposal.

The finance should come from Mines and Minerals Department of each State, each mine owner contributing money for his own mine, and the Social Forestry Wing will spend the money.

To execute this work with success integrated management plan has to be prepared for a cluster of mines in an area besides management plan for each mine separately.

Various states have already formed state and district level coordination committees on Social Forestry matters. Panchayats have also been involved in various activities. Extension staff have to be carefully selected. Some states have already started establishing biomass garden, clonal bank/progeny orchard, agroforestry, socioeconomic services, linkage with various rural extension services.

Finally the authors want to stress that people have to be adequately educated to realise their position and develop desire to have a better quality of life as meant in the following:

A Chinese adage

- If your plan for one year plant rice.
- If your plan for ten years plant trees.
- If your plan for 100 years educate mankind.

In words of Swami Vivekananda:

"All the wealth of the world cannot help one little Indian village if the people are not taught to help themselves.

8

Foresters are Pioneer Ecologists

— Foresters were Pioneers in the Application of Ecological Principles and Theories

Analysis of the history of forest administration shows that several pioneer foresters on their very first visits to the country recognized the need for conservation and preservation of the forests in the country. As a result of mass-scale deforestation and unscientific management of forest resources, wide-scale soil erosion and climatic changes have occurred. They felt the urgent need for scientific management of existing forest resources and also extending forest areas by creation of plantations over the waste lands.

The authors have summarized the observations of the visionary forester Ribbentrop, who realized the calamities that the country might face and recommended various ecological and administrative measures to solve the problems.

Foresters are Pioneer Ecologists

— Foresters were Pioneers in the Application of Ecological Principles and Theories

Ribbentrop, the German forester who also held the post of Inspector General of Forests, Govt. of India, was a visionary.

He made an assessment of the land use situation in some parts of India about which he recorded in his book, "Forest History of British India (1906)". It was extremely a creditable job as in those days communication and very many other basic amenities of movement and stay were not developed.

To testify the authors' decoration of the epithet 'Ecologist' to foresters a few quotations from Ribbentrop's book has been thought to be very relevant, some of which are presented below:

- critical situation might arise due to overfelling of trees,
- rampant grazing and fire that are bound to destroy all regeneration on forest floor,
- soil erosion effects silting of river beds,
- felling of trees directly affects the climate of a place,
- and several other related issues.

Forest Denudation vis-a-vis Rainfall

With regard to the actual decrease of rainfall consequent on the destruction of forests, Major-General Fisher, R.E., an old resident of Bellary and Remandrug, supplies the most interesting information in the following note.

"I arrived in the Bellary District in June 1856 and visited Ramandrug at once; the hills were then covered with a good strong jungle; there was always heavy cloud during the night resting on the hills and for the great part of the day; rain fell during the southwest monsoon constantly and frequently; during the northeast monsoon it was much lighter; in the months of March, April and May, the monsoon showers were usually very heavy and accompanied with much thunder and lightning. The average rainfall we calculated was then 45 inches in the year; all the springs about the hills ran abundantly throughout the year, and the Nareehulla, the main feeder of the Darojee tank, with all its tributaries, had water running in them all through the year. The climate of the Drug during the monsoons and cold weather was quite cold enough to make fires very necessary, although its elevation is not more than 3,300 feet above sea-level. The water supply was most abundant during the whole of the hot weather, and the tank was almost always full, surplusing very largely during the southwest monsoon."

"These observations refer to the years 1856 up to 1864 inclusive, when I left the Bellary District and did not visit the Drug again till January, 1879. I found everything changed; the jungle has been almost entirely destroyed; the rainfall is most precarious, and certainly not so much as 24 inches in the year; the tank has not filled for the last three years, and is generally 10 or 12 feet below full tank level; the springs are almost always dry, dribbling only at the best; the climate is so changed that in the cold weather it is hardly necessary to shut the doors and windows; except for the high wind and slight mists of the southwest monsoon, it would not be necessary to close the house at all. The main feeder of the Darojee tank dries up altogether by the end of February, and all its tributaries have no water in them."

Mr. Macartney, the Agent of the Sandur State in Bellary District, Madras Presidency, also maintains that the rainfall within the last ten years has become lighter and more irregular with the increased destruction of forests by woodcutters and charcoal-burners, and the indiscriminate grazing of cattle, sheep and goats. Mr. Macartney speaks from an experience extending over 22 years, and supports his observations by the following facts.

"In the first decade of my residence here, the tank near my house used to be regularly filled every year and to be running over for several weeks at a time. Latterly, though, it has accumulated an immense amount of silt, and is now consequently of diminished capacity; it rarely fills. The same remarks apply equally to the Ramandrug tank and to that of Singankeni."

The Rushikulya, a river in the Ganjam District, Madras Presidency, is formed by two main branches. One of these, coming from the well-wooded hills of Surada and Pandakol, carries water for nine or ten months in the year; whereas the other, the Mahanadi, taking its rise in the much more open country of Gumsur and Chokapad, is dry for nearly eight months.

A difference of 10 per cent in the average rainfall, combined with a more equal distribution, especially over the drier months, would suffice to bring about the historical changes already noted.

Ebemayer's exhaustive experiments have shown that the mean annual temperature in a closed forest is 10 per cent less than in the open. In both cases the measurements were taken at 5 feet from the ground. The difference is greatest in summer, and, consequently, to us in India of much greater importance than to the inhabitants of more northern latitudes.

That when a much greater part of India was covered with forests the climate of the country was on this account different seems equally indisputable. Fa-Hian, the great Chinese traveller in India in the 4th century (A.D.), says, in describing the country, that its temperature was neither cold nor hot. This, as already stated, is corroborated by the ruins of an old civilization in many localities where this could not now exist without large works of artificial irrigation. It may, I think, be assumed that the same state of things would re-establish itself if a considerable proportion of the country were again brought under dense forests, which is physically possible; but that it could be effected to the extent desirable by any measures of forest conservancy and that any Government could under existing circumstances apply all out effort.

During the earlier period of the present administration, shifting cultivation (*kumri*) was practised to a large extent in the Central Provinces, and several thousand of square miles were thereby laid barren year after year. Early in the seventies this method of cultivation was stopped and extensive growth of young forests sprang up in its place. It is true the whole area is grazed over to a considerable extent and cannot therefore exercise that influence which it would exert under more favourable conditions, but it would appear that the influence has nevertheless been beneficial.

Local Influence of Forests on the Fertility of Country

There is no divergence of opinion regarding the more local effects of forest in protecting the soil and regulating both surface and subsoil drainage. In an open country the greater portion of the rain water runs off, and on a steep decline with great rapidity and force, carrying the fertile soil along with it and cutting deep ravines and gorges.

As long as the water carries fertile soil, the underlying country is often benefited by inundation, especially if this happens during the right time of the year. Soon, however, it is found that the good surface soil has disappeared and the rushing torrents carry unproductive sand, stones, or clay, and smiling fields are covered with layers of dead soil unfit to produce a crop for years. The blessing of the proximity of chains of hills and mountains is turned to a curse, and in most instances by man's own action.

Almost every book treating of this subject teems with examples regarding the baneful results of deforestation which might be cited; but we need not go to books, for Nature's own proofs are plentiful in this country.

Denudation of Hill Forests

The Hoshiarpur chaôs situated in a rich agricultural country the north of India, are – owing to the fact that the hills from which they spring are to a great extent composed of very friable and sterile sandstone – one of the most marked examples of injury which may result from the denudation caused by the destruction of forest growth. Year by year considerable additional areas are covered by unproductive sand, causing an incalculable loss to the country, which may be gauged by the fact that the loss in land-revenue alone since the last settlement was made said to amount of Rs. 90,000 per annum. Sufficient proof exist that the hills in question were once densely wooded, and that the destructive torrents did not then exist. Afforestation would cure the evil and form a monument of Indian forestry that could easily be equalled.

The hill range between Mahasu and Fagu, in the vicinity of Simla, well known to all residents, was, as late as 1868, covered by a magnificent forests of Spruce, Silver-Fir, Blue-Pine, Deodar and Oaks. The ground was subsequently cleared for potato fields, and some fine crops of excellent potatoes were gathered. Now, however, the soil has been washed down into the ravines, the fields have to a great extent disappeared, and the barren hillside is cut up by the dry stony beds of Alpine torrents. These are but a few of the many examples in the Himalayas and sub-Himalayas where cause and effect may both be traced within the time of our occupation of India. Results of former and more ancient causes are discernible wherever the eye is cast.

Proofs of the lamentable consequences of deforestation in other parts of India are not wanting.

Denudation in the Ceded Districts of the Madras Presidency yearly covers with sand fresh areas of formerly culturable land. Even in the comparatively well-wooded valley of Pullampet, Cuddapah District, the consequences of denudation (due to indiscriminate cutting and over-grazing) are becoming very marked.

In Kanara numerous instances are reported where spice gardens near the Ghats have had to be abandoned on account of the destruction of the forests in the vicinity, and even within the once moist and cool valleys of the Sirsi and Siddapur ranges, gardens were deserted soon after the hillsides had been cleared of forest growth.

Similar observations have been made in Assam, where the surface-soil in many tea-gardens has been washed away, and where the yield has consequently dwindled down to next to nothing.

The Ratnagiri District, in the Bombay Presidency, is almost bare up to the crest of the Ghats, and here, Sir Dietrich Brandis says, the effects of denudation have shown themselves in this way.

"There are four principal streams in the district, which, rising in the Ghat mountains, run a short course to the sea, all of which were formerly navigable and important for the trade of the country. For small boats they are still navigable, but they are gradually silting up because the hills at their headwaters have become denuded of forests."

The slopes on the west coast of the Bombay Presidency were once even during the earlier days of British occupation, clad with magnificent and most valuable and extensive Teak forests; but these have long since disappeared.

The denudation of the Deccan highlands and the Eastern Ghats has resulted in the gradual silting up of rivers. When the English, French and Dutch first made settlements on the Coromandel Coast, they were able to take ships up the rivers Godavari and Kistna, Narasapur (English) and Yanaon (French) on the Godavari, though now only approachable by small native crafts at high tide, were once the chief ports for that part of the coast. At Masulipatam the Dutch ships used to come close up to the fort, but now even native vessels of small draught have to anchor 5 miles out in the roads.

It is also said that a hundred years ago, when the Ganjam town and fort were places of considerable importance on the Madras coast, small seagoing vessels used to cross the bar of the river and lie at anchor opposite the fort – a place where there is now barely 2 feet of water in the dry season. Dr. Gibson, the first Conservator of Forests in the Bombay Presidency (1847), give in one of his earlier reports, a list of creeks and rivers of the Malabar Coast where ships used to ride at anchor, and which had silted.

In the Trichinopoly District, Madras Presidency, considerable planting operations have been carried on during the past ten or twelve years along the banks of the Cauvery. Observations prove that whereas wells from 6 to 10 feet deep in the plantations are well supplied with water throughout the hot weather, even when the river and neighbouring channels are quite dry, wells 15 feet deep on unplanted land in the neighbourhood, on the same level and otherwise similarly situated, are quite dry throughout the hot months.

The question is of even greater importance for India than in Europe, and no better arrangements with regard to this can be brought forward than those used by Sir Dietrich Brandis in his pamphlet entitled "Progress of Forestry in India".

"In a large portion of India the crops depend either partially or wholly upon irrigation, and the water is derived from tanks, wells, or rivers. The tanks are water reservoirs of various extent, generally constructed by damming up a stream or river in a convenient place; but there are also smaller storage tanks, which are fed only by the surface drainage flowing direct from the catchment area. Tanks of this latter description would store the largest proportion possible of the water coming from the catchment area if that area were made impermeable to the rain which falls upon it. In such cases trees and forests upon the catchment area are injurious, and several instances have been observed where such tanks have ceased to be filled since the forest on their catchment area has become dense and heavy. It is different with the larger tanks, which are fed by springs and streams. These benefit largely by thick forest growth upon their catchment area. The area irrigated from wells in the different parts of India is very large. Thus, in the Madras Presidency two million acres are irrigated from wells, while three million acres are irrigated from rivers and tanks. There is no reason to believe that forests growth on level ground has the effect of raising the subsoil water level, which is tapped by the wells; but wells are frequently dug at the bottom of a valley or near the bed of a stream, and in such places there is ground for believing that the underground water stratum which is tapped by the wells will be supplied more plentifully, and that the supply will be maintained longer during the dry season, if the hills which surround the valley are clothed with dense forest growth. Wells of this description are numerous in Ajmer and Merwara and one of the objects for forming the forest reserves in those districts was to improve the water-supply in these wells."

The rivers which feed the canals used for irrigation are of two classes. The Ganges and Indus, and their tributaries, are fed by the snow which falls on the Himalaya mountains and by the plentiful summer rains of the monsoon. The water-supply of these large rivers cannot materially be affected by the small forest area which it may be possible to place under good management on those mountains, but between these big streams a large number of smaller rivers exist, the drainage area of which is situated in the lower hills and which join the main stream below their debouchment from the mountainous region. Of the other rivers which are used for irrigation, the most important are the Sone in Bengal, the Mahanadi in Orissa, and the Godavari, Kistna and Cauvery in the Madras Presidency. These rivers and their feeders rise in the hills of Southern and Central India which derive their water-supply chiefly from the summer rains of the southwest monsoon. In regard to these rivers there is good ground for believing that their water-supply is largely affected by the forest growth upon their catchment area, and that in some cases denudation has already done great damage in this respect.

The Tambraparni, in the Tinnevelly District, is one of the smaller rivers used for irrigation in India. Its catchment area is 1,739 square miles, of which 1,389 are in the plains and 350 square miles in the hills on the eastern slopes of the Ghats. The river irrigates the large area of 170,000 acres or 265 square miles of rice-fields, more than one-third of which bears two crops. It is a beautiful sight to see this large expanse of brilliantly green fields at a time when the country around is parched and barren. This river carries down into the plains a very large proportion of the rain which falls upon its catchment area, and there is good ground for believing that if the forests, which fortunately are dense and extensive near its head waters, were cleared, a larger proportion of the water would be lost by evaporation, and that the supply, which now flows almost uniformly during the greater part of the year, would come down in sudden rushes after each heavy fall of rain, and would be more irregular.

The increased suddenness and violence of floods in hill-streams, as an immediate consequence of the destruction of forests on their catchment areas, have been noticed in many parts of country. The most remarkable observations in this respect have been made by the officers on the Jumna Canal, with regard to the streams opening into this waterway. From Cuddapah District, in the Madras Presidency, similar reports have been received, and it is stated that the floods in the Gandaleru, Ganjana, amd Moshtiyem hill-streams, which take their rise in the Seshachellam hills, have increased in suddenness and violence since the density of the forests on their catchment areas has decreased.

The periodically, recurring breaches in the railway embankments, especially those of July 1866, are, there is good reason to believe, due to the denudation of the Siwaliks and other sub-Himalayan hills.

In the same way as the hurtful effect of deforestation on soil and drainage is chiefly felt when the ground is hilly and in a degree proportionate to its steepness, the beneficial effect of the action of forests in this respect is greatest under similar circumstances. It would again be easy to quote numerous examples based on observations made in other countries, and it would be more excusable, as results due to strict conservancy are much rarer in India than warning examples; for only completely stocked forests which are free from conflagration, and the ground of which is covered with dead leaves, twigs and vegetable mould, have an appreciable effect as regards protection of the soil and regulation of subsoil drainage. Open forests, in which the grass and vegetables debris are consumed by annual fires, or from which the mould and dead leaves and twigs are constantly removed by "*sur*" or "*rab*" collectors, have no effect in this respect.

In Chamba, Kulu, Bashahar, and in the immediate vicinity of Simla, the results of strict conservancy in protecting the soil and in regulating the water-supply are clearly discernible to ever one who has revisited these localities after an absence of ten or twelve years; but these observations are perhaps too general to carry conviction.

The beneficial effect of forests properly protected was, however, very marked in the following case. In 1878 a road was built from Ellichpur in Berar into the heart of the Melghat forests, which had been protected against fire for ten years, and had consequently become in most parts completely stocked with a heavy layer of old leaves and grass on the ground. The road runs partly through the forest thus protected and partly through the open or unreserved forests, where no protection had been attempted, and the Executive Engineer who constructed the road wrote as follows regarding the effecting protection upon floods in that part of the country.

"During the late heavy rain, *viz.* 4 inches in 24 hours in the beginning of June 1987, I had four bridges in construction in the Bairagarh Reserve on the Ellichpur-Pili road *viz.* Nos. 18, 19, 22 and 23. The foundations of these bridges were not finished, and I expected much damage to them by the early rains. Much to my surprise I found that when the rain was finished none of the *nalas* over which these bridges are being constructed had been in flood and no damage had been done to the foundations. In the unreserved forests, between Sirasban and Ellichpur, all the *nalas* had been in flood, and I attribute the escape of the foundations of the bridges in the reserve to the rain having been absorbed by the old fallen leaves and grass, thus showing the protective power of forests to bridges in preventing sudden heavy floods."

These extracts from the work of Ribbentrop depicts a very realistic picture of central and southern India more than a hundred years back and how efficiently the foresters made the assessment.

Ecological Activities of the Forests were

Some factual examples of forester's ecological activities may be summarized as follows:

- To save the forests from utter depletion from fire they adopted fire protection methods.
- To augment regeneration the foresters manipulated and controlled firing of regeneration sites.
- To save the forests from pilferage and eventual destruction fuel reserves and village forests were created and judicious felling methods were prescribed.
- The foresters compiled flora of various regions to have a clear knowledge about forest resources, the evolutionary trend and species richness and geographical elements.
- Linear valuation survey plots were established to measure species, their diameter classes, their regeneration and growth, etc.
- They checked over the unregulated felling of private agencies in various forests.
- In 1894, the foresters enacted a Forest Policy for the maintenance of forest cover and amelioration of climate.
- The foresters started artificial regeneration of valuable species since 1840 (Nilambur) but formally since 1896.
- The foresters stopped over-exploitation of forests to save soil from erosion.
- The foresters marked erodable, friable steep hilly areas and put them under "Protection Working Circle".

- The foresters created preservation plots for protection of useful species and record succession of species.
- The foresters created Game reserves, Wildlife sanctuaries, National parks, Biosphere Reserves, etc. for protection and conservation of flora and fauna.
- After creation of plantations the foresters took care and looked after the growing crop by manipulation of congestion and density for the benefit of the crop for maximum yield and getting intermediate yield.
- The foresters worked intensively and successfully marked the
  - indicator species
  - plant succession pattern
  - climax forest formation
  - pure plantation and their evil effect
  - healthy mixture of several species
  - regeneration by manipulating the overhead canopy
  - proper management code for upkeep of departmental elephants
  - plants as pollution indicators.
- The foresters also determined the following:
  - Forest types existing all over the country.
  - Various associations, consociations, etc.
  - Various forest tiers and their ecological role.
  - Grassland types vis-a-vis tree regeneration.
- Plant biomass vis-a-vis herbivore conception were those of the foresters.
- Succession of vegetation on creeks, river banks, land slides, fire burnt sites.
- Forest ground flora and wildlife relationship.

The authors sincerely desire that some competent forester should take up the job of recording the colossal ecological changes that occurred all over Indian subcontinent in the last 100 years (1905-2005), i.e. after the realistic and educative observation of Ribbentrop. Such observations will surely testify to severe land-use changes in spite of sincere efforts of the foresters.

9

Management of Wildlife Sanctuaries in Aquatic Sites

— Forester's Achievements

*A*quatic sites in India are a dwindling resource. Various anthropogenic pressures have caused reduction of habitat and biotal resources.

The authors firmly believe that all the aquatic areas under wildlife sanctuaries and National Parks are scientifically administered by the Forest Departments. But there are problems and various facets of management.

They also feel that all the water bodies of the country i.e. ponds, lakes, nallahas, canals and rivers should strictly and intensively be reclaimed and controlled. In the dry and arid areas more waterbodies need to be created.

The authors have drawn an outline on all these facets and stressed that the forest personnel of the protected areas must undergo refreshers' courses in advanced management of waterbodies.

Management of Wildlife Sanctuaries in Aquatic Sites
— Forester's Achievements

Natural Water Bodies

Natural water bodies are a part of vast aquatic ecosystem that plays a vital role in the biogeochemical cycle of the entire earth which effect the composition of the atmosphere, the climate and hydrological cycle. Climate determines many of the critical characters of aquatic ecosystem. The most important aquatic ecosystem in global biogeochemical interaction are the oceans due to their unlimited and immeasurable size. However, little they are, the inland water sites on the earth cannot be underestimated for without it, the earth would be devoid of all terrestrial and fresh water species including man. Water is one of the basic resources of civilization and forms the life blood of all living organisms of the earth. The aquatic environment is in a continuous state of change, with fluctuations following a daily and seasonal cycle on which climatic events of rainfall are superimposed.

In India the situation has assumed an alarming state owing to siltation, pollution, extension of agriculture, universal creation of embankments, heavy anthropogenic pressure, settlement of displaced people and unscientific use of many such situations.

A search for appropriate scientific literatures on waterbodies (ponds, reservoirs, ox-bow lakes, canals, haors, beels, jheels and rivers) in relation to various resources like fish, birds, reptiles turtles and tortoises, amphibians, zoo and phytoplanktens, filamentous algae, periphytons, benthic biota, bottom organisms, physicochemical properties of water and soil, macrophytes, etc. clearly reveals that voluminous work has been done in the universities, State Fishery Organizations and various fishery research institutes of Govt. of India and ICAR in particular. The CIFRI (CICFRI) have already done, voluminous work on most of the aforesaid parameters. In this connection mention may be made of the journals *viz., Journal of Indian Fisheries Society of India* (J.I.F.S.I.), various proceedings of Zoological Society, Fishing Chimes, Annual Reports, Workshop papers held in such Research Institutes and Organizations of urban areas have been damaged and some lost. Yet one happy note is that the Ministry of Environment and Forests, Govt. of India in their lists of 'protected areas of the country' have included some of the best and now endangered aquatic sites of country (Loktak, Kaziranga, Sundarbans, Keiladeo Ghana, Ranthambhore, Kolleru, Pulicut, Chilka and many others) which would receive long time protection and these will hold the biodiversity resources. On the other hand ponds, lakes and reservoirs which would be brought under fishery projects may to some extent eliminate some of the macrophytes and biodiversity of these waterbodies would be little disturbed.

Lakes, Flood Plains and other Waterbodies

Lakes are another type of waterbodies with comparatively a stable environment. There is less of turbidity more development of phytoplankton and by the process of photosynthesis, they bring about the production of organic nutrients and oxygen.

Flood plains areas are subject to seasonal flooding permanent or semipermanent areas of standing water may be locked in oxbow lakes and other depressions after the flood waters have receded. Such areas have been a vital dry season refuges for fish and wetland animals. These areas also sustain a heavy population of livestock.

Various facets of wetland which come under the mandate of Ramsar Convention may be appropriate but some of these are not applicable to Indian condition. Of the seven components such as estuaries, peatlands, open coasts, flood plains, freshwater marshes, lakes and swamp forests, this author does not consider that estuaries, peatlands and swamp forests should be included in the list of wetland in India. Estuaries, in India, occur over extensive areas and are a special feature of a very specialized habitat. Similar peatlands cannot be properly identified in this country, while the swamp forests are absent except for a few types (littoral swamp and tropical freshwater swamp) identified by Champion and Seth (1968).

Management of Aquatic Sites (Foresters must have the basic studies)

Natural water bodies which are one of the few biotal resources sinks are facing extinction due to erosion, pollution, over-exploitation of plants and animals resources, poaching, human settlement expansion, etc. In recent years the scientists of inland Fisheries Institute and Universities have brought to light the horrors of such depletion in published papers into various scientific journals.

Due to such perilous impact the following biotal resources have faced depletion involving:

- Macrophytes, phytoplankton, zooplankton, benthic biota, periphyton, etc.
- Very many species of fish fauna reducing the source of protein food.
- The population of dolphin, porpoises, turtles and tortoises, crocodiles and several other species.
- Both resident and migratory bird population.

The vast and most useful aquatic biotal resources will now find a safe home only in the protected areas of Forest Department in all the States duly managed by the foresters.

This emphasize the necessity of the managers of "Protected Areas" to undergo a thorough scientific training on the management of such Water bodies to ensure survival of all the biotal resources.

(Refer the Chart)

The author's treatise viz., "Wetland Ecology" (1996) which records various aspects of Indian water bodies is very relevant to the present issue. He feels that Indian Foresters who are sole custodians of these water bodies inside reserve forests must undergo an intensive management course of training on aquatic floral and faunal biology a broad outline of which is drawn in this Chapter. They need to have a thorough knowledge of water chemistry. A small laboratory is an essential requirement for this study. Various Universities besides the State and Central Fisheries have their laboratories, but for prompt remedy of various problems a own small laboratory in each State is a must under the control of Forest Department. An adage very relevant to quote here that –

"Man, man is the devil,
The source of all evil."

Of all living organism of earth, only man has seriously threatened the delicate natural balance existing among these four groups. He is now increasing his threats to the point that he threatens not only the existence of non-human organisms but also his own, for man is the only organism that pollutes and also manipulates his environment to serve his own needs and desires.

Human Population and Environmental Exploitation

Many of the current ecological problems of the environment are really based on rather simple ecological principles. The balance of nature remains a balance only if there exists a proper give and take relationship along the following four groups:

WILDLIFE MANAGER'S NEED TO KEEP INFORMATION ABOUT VARIOUS FACETS OF WATER BODIES

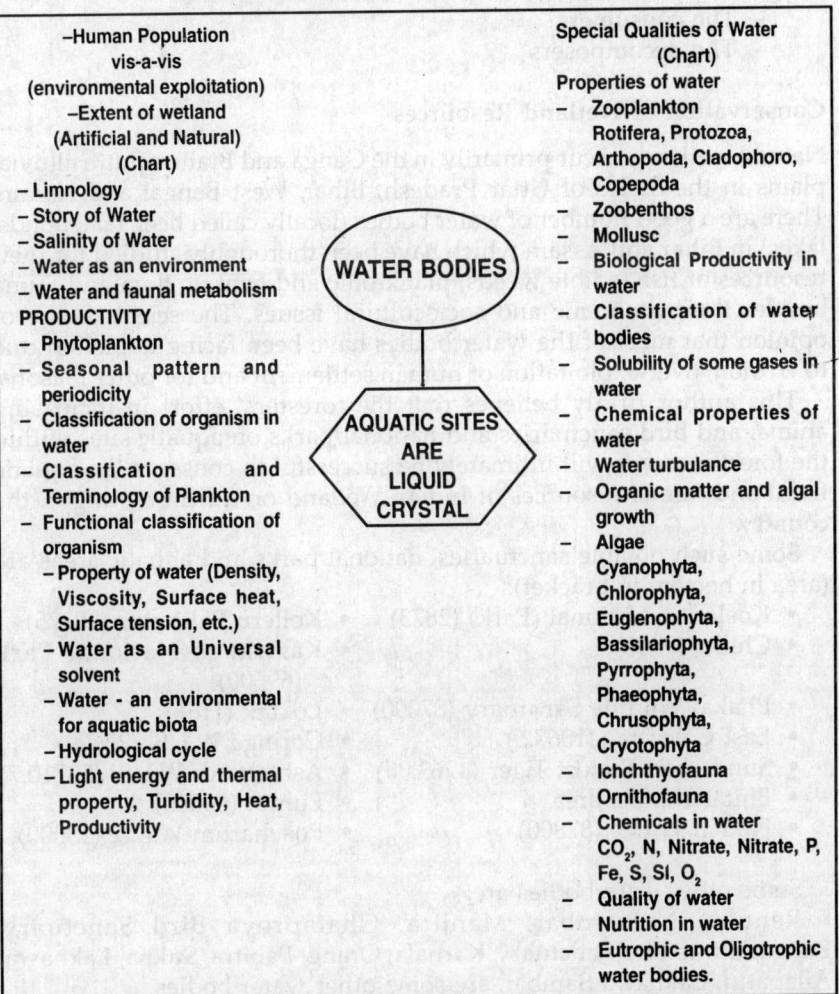

–Human Population
vis-a-vis
(environmental exploitation)
   –Extent of wetland
   (Artificial and Natural)
      (Chart)
– Limnology
– Story of Water
– Salinity of Water
– Water as an environment
– Water and faunal metabolism
PRODUCTIVITY
– Phytoplankton
– Seasonal pattern and periodicity
– Classification of organism in water
– Classification and Terminology of Plankton
– Functional classification of organism
   – Property of water (Density, Viscosity, Surface heat, Surface tension, etc.)
   – Water as an Universal solvent
   – Water - an environmental for aquatic biota
   – Hydrological cycle
   – Light energy and thermal property, Turbidity, Heat, Productivity

WATER BODIES

AQUATIC SITES ARE LIQUID CRYSTAL

Special Qualities of Water
(Chart)
Properties of water
– Zooplankton
   Rotifera, Protozoa, Arthopoda, Cladophoro, Copepoda
– Zoobenthos
– Mollusc
– Biological Productivity in water
– Classification of water bodies
– Solubility of some gases in water
– Chemical properties of water
– Water turbulance
– Organic matter and algal growth
– Algae
   Cyanophyta, Chlorophyta, Euglenophyta, Bassilariophyta, Pyrrophyta, Phaeophyta, Chrusophyta, Cryotophyta
– Ichchthyofauna
– Ornithofauna
– Chemicals in water
   CO_2, N, Nitrate, Nitrate, P, Fe, S, Si, O_2
– Quality of water
– Nutrition in water
– Eutrophic and Oligotrophic water bodies.

- The producers that is, the autotrophs or photosynthetic organism that capture solar energy and use to produce food stuffs, directly or indirectly, for all forms of life.
- The consumers that is the heterotrophic or organic living either directly or indirectly off food product provided by the autotrophs.
- The decomposers are primarily bacteria and fungi.
- The inorganic world that is, the nitrates, phosphates, carbonates, etc. all of which are needed in balanced amount as nutritional supplement by both producers and consumers.

- The inorganic world
- The producers
- The consumers
- The decomposers

Conservation of Wetland Resources

Natural wetlands occur primarily in the Ganga and Brahmaputra alluvial plains in the States of Uttar Pradesh, Bihar, West Bengal and Assam. There are a good number of water bodies (locally called beel, taal, ponds, lakes) in Bihar and Assam which have been thoroughly studied for their resources of fish, edible weeds, planktonic and benthic flora and fauna besides their economic and sociocultural issues. The scientists are of opinion that most of the water bodies have been facing depletion due to erosion, over-exploitation of human settlement and for other reasons.

This author firmly believes that the foresters' effort in increasing animal and bird sanctuaries and national parks on aquatic sites within the forest reserves will ultimately be successful to conserve the aquatic floral and faunal resources of Indian Wetland on a firm footing in the country.

Some such notable sanctuaries, national parks and aquatic areas are (area in hectare in bracket):

- Koelodeo National (Park) (2873)
- Chilka (9400)
- Kolleru Palicanry (28375)
- Kaziranga National Park (69600)
- Phakal Wildlife Sanctuary (87000)
- East Calcutta (11067.5)
- Sunderban Project Tiger (4263.00)
- Bhitar Kanika area
- Pakhal W.L.S. (87800)
- Loktak (1166)
- Coringa W.L.S. (23600)
- Ashtamudi W.L.S. (50710.7)
- Puhiot (60000)
- Poacharum W.L.S. (13000)

Some other waterbodies are:

Renuka, Nalsarobar, Manjira, Ghataprova Bird Sanctuary, Ranganthitoo Bird Sanctuary, Karnala, Orang, Pabitra, Sukna, Lakhawa, Ailapattu, Lakhawa Sambar, are some other water bodies.

Some Preliminary Notes on Water and Water Bodies

Limnology (Limnos : means lake, pool or swamp. Logo[s] means study).

Limnology is the study of fresh or saline waters which are contained within continental boundaries (Inland water).

In a broad term Limnology is the study of functional relationships and productivity of biotic communities as they are affected by the dynamics of physical, chemical and biotic environmental parameters (Wetzel). It is the science of inland waters viewed as ecosystems together with their structure materials and energy balances.

STATEWISE WATERBODIES INSIDE FOREST AREA AND OUTSIDE
(Area in ha.)

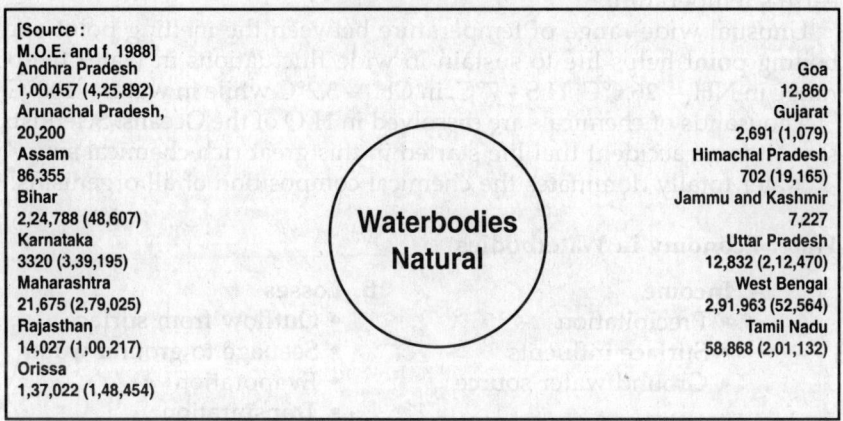

[Source :
M.O.E. and f, 1988]

Andhra Pradesh	Goa
1,00,457 (4,25,892)	12,860
Arunachal Pradesh,	Gujarat
20,200	2,691 (1,079)
Assam	Himachal Pradesh
86,355	702 (19,165)
Bihar	Jammu and Kashmir
2,24,788 (48,607)	7,227
Karnataka	Uttar Pradesh
3320 (3,39,195)	12,832 (2,12,470)
Maharashtra	West Bengal
21,675 (2,79,025)	2,91,963 (52,564)
Rajasthan	Tamil Nadu
14,027 (1,00,217)	58,868 (2,01,132)
Orissa	
1,37,022 (1,48,454)	

Waterbodies Natural

Biological productivity of anti body of water is the end result of the interaction of the organisms present with the surrounding environment. Biological productivity includes its qualitative and quantitative features and its actual and potential aspects.

The foresters are aware of the pros and cons of the subject related to aquatic site management on a broad base.

Story of Water (Waterbodies)

– What is the most common thing in the world?
– What is the most unusual chemical in the world?
– What is the most essential thing for Life?

The only answer water which is life.

• Water is absolutely essential to very form of life.
• Water covers over 70% of earth's surface
• You and I are also 70% water ... Do you believe?
• Water is a liquid, a solid (ice) and a gas (vapour).

Unusual Properties of Water

When any liquid cools – it contracts – becomes denser. When water cools below 4°C, it starts expanding again. So, at 0°C (ice), it is less dense. Ice is only 9/10 as heavy as same volume of water. It expands on freezing.

It prevents entire water column in a lake or ocean to get frozen. Surface water gets frozen while denser water at 4°C sinks down and sustains animal life.

Specific heat of water is very high. It gains/loses large amount of heat without much change of water temperature. This helps in regulating earth's temperature.

Unusual wide range of temperature between the melting point and boiling point helps life to sustain in wide fluctuations in water temp. range in NH_3 – 26.6°C; H_2S – 7°C; in CH_4 – 3.2°C while in water – 100°C.

Thousands of chemicals are dissolved in H_2O of the Oceans. Scientists say, "It is no accident that life started in this great rich chemical soup."

Water totally dominates the chemical composition of all organisms.

Water Economy in Waterbodies

A. Income
- Precipitation
- Surface influents
- Ground water source

B. Losses
- Outflow from surface
- Seepage to ground water
- Evaporation
- Transpiration

The Hydrological Cycle

The cycle consists of three principal phases:

- Evaporation
- Precipitation
- Ground water run off

Each phase involves transport, temporary storage and a change in the state of the water.

Water as an Environment

Water has two-fold effect on living organisms in Waterbodies :

- Through its physical properties – Medium in which organisms live.
- Through its chemical properties – supply nutrients for primary production.

Physical Properties

The properties are:

Specific heat, latent heat, specific gravity or density, pressure, compressibility, buoyancy, viscosity and surface tension.

Density

Water is 775 times denser than air. Aquatic organisms do not have to carry their body-weight-thus save energy, also reduction of supporting tissues.

Greater density exerts a marked buoyancy on organisms. Greater the density, greater the buoyant force. The density increases with decrease of temperature and reaches maximum at 4°C after which it decreases. Sp. gravity of pure ice at 0°C is 0.9168.

Density varies with : (a) Temperature, (b) pressure, (c) dissolved substances, and (d) suspended particles.

Buoyancy is a direct outcome of density and varies with the same factors.

Pressure

Water is a heavy substance. 1 c.c. weighs 1 g at 4°C. It increases with depth. Pressure increases one atmosphere with 10 m depth objects may collapse under heavy pressure of water – called Implosion.

Compressibility

Water is virtually incompressible.

Viscosity / Mobility

Water is an exceedingly mobile liquid. But it has internal friction. Viscosity is a measure of a liquid's resistance to flow. Viscosity of water is 100 times greater than air. It falls with the increase of water temperature.

	Temp. °C	% of Viscosity	
	0	100.0	
	5	84.9	
Water has low	10	70.0	Dissolved salts
viscosity as	15	63.3	increases viscos-
compared to	20	56.1	ity of water.
other liquids.	25	49.8	
	30	44.6	

Changes in the viscosity of pure water are due to changes in temperature.

Surface Tension

Water has high surface tension. It occurs due to unbalanced attraction between molecules. Since the surface molecules are attracted on one side only upward attraction is lacking since there are no water molecules above them. This property is important in the maintenance of protoplasmic form and movement. It acts at the air-water interface and forms a special habitat for organisms adapted to living in surface film (Neuston) organic surfactants (forming or wetting agents) produced by aquatic plants and animals reduce surface tension. S.T. is also reduced by dissolved salts. Water striders (insects) take advantage of surface tension and skate rapidly on the water surface.

Special Quality of Water

The molecular structure allows weak hydrogen bonding between hydrogen and oxygen atoms in adjacent molecules, producing a matrix known as a liquid crystals.

The maintenance of this liquids crystal gives water most of its universal properties.

If water behaved at natural environmental temperatures as do H_2S, NH_3 or CH_4, it would be present only as vapour. In fact, water and mercury are the only inorganic liquids that exists at earth's surface under N.T.P (normal temp. and pressure).

The complex bonding holds the water together as a liquid to a much higher temperature than compounds H_2, NH_3 or HF.

Water as a universal solvent, transports through Biological systems the gases, minerals and dissolved organic components that drive life's machinery.

Water's unusual properties : Such as changes in its density at boiling and melting points, viscosity, surface tension, specific heat, high dielectric constant and others have made water "LIFE SAVING COMPOUND".

Water and Aquatic Faunal Metabolism

Water exhibits some unique properties which derive from the structure of the water molecule and its tendencies to form aggregates.

Water molecules are strong dipoles and experience a strong attraction for each other and associate to form hyper-molecular linear, areal and spherical clusters. This results in a complex electrostatic system of hydrogen bridge [Bonding] formation. Each water molecule is connected to four adjoining molecules by means of hydrogen bridges forming roughly a tetrahedral arrangement. This makes water a most active chemical compound of extensive significance in the metabolism of living organisms.

Biological Productivity

Biological productivity of any body of water is the end result of the interaction of the organisms present with the surrounding environment. Each organism possesses inherent characteristics and abilities which enable it to exist and reproduce. This is known as the biotic potential of the organism. However, conditions in the environment affect and restrict and organism and there is always constant struggle between biotic potential of the organism with the factors in the environment. Also there is a constant struggle among the organisms themselves (intra and inter-specific struggle).

Some physical factors of waterbodies influence biological productivity (B.P):

Under strictly comparable conditions, the greater the length of the shoreline, the greater the biological productivity of the waterbodies. The increased irregularities of shore line results in:

1. Greater contacts of water with soil – and more nutrients are released
2. Increased areas of protected bays and caves. Protected shoal permits :
   (a) Growth of rooted plants; (b) accumulation of organic bottom materials; and (c) presence of much larger and more diversified fauna.
3. Increased shallow areas for growth of rooted vegetation.
4. Greater diversification of bottom and margin condition.
5. Reduction of the amount of exposed, wave-swept shoal; and
6. Increased opportunity for extensive, close super position of the photosynthetic zone upon the decomposition zone.

According to their mode of nourishment, the organisms present in a body of water may be classified as under.

Autotrops (self-nourishing organism) : Those organism that are able to prepare organic matter from inorganic with the help of solar energy, e.g. chlorophyll-bearing plants.

Heterotrophs – Those organism that are unable to prepare their food and are dependent upon autotrophs for food directly/e.g. animal.

What are producers, consumers, decomposers? And what are Inorganic instalment.

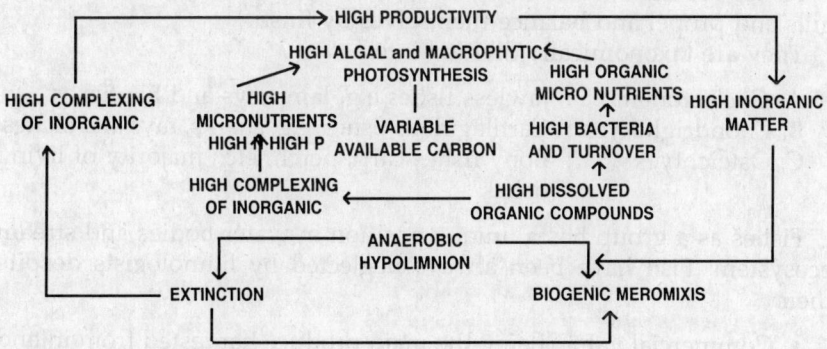

Major inorganic and organic interactions influencing the metabolism of phytoplankton of eutrophic water bodies.

Some Important Characters of Eutrophic Water Bodies

- Relatively Shallow; Deep cold water minimal or absent.
- Organic materials on bottom and in suspension abundant.
- Electrolytes variable, often high; Ca, P and N abundant; humic materials slight.
- Dissolved oxygen in deeper stratified water bodies minimal or absent in hypolimnion.
- Larger aquatic plants abundant.
- Plankton quantitatively abundant, quality variable, water blooms common; blue green algae and diatom predominant.
- Profoundal fauna in deeper water bodies poor in species and quantity in hypolimnion, chironomus type.

Table is given on page 171.

Fish and Wildlife

Fish has not been considered as a wildlife group in the definition of Wildlife Act in India. The United States of America however had since long past considered fish as a part of wildlife. The author firmly believes that the definition of wildlife suffers a wide lacuna owing to this disregard of Ichthyofauna, which are an integral part of aquatic environment, provides food to birds, animals and man and form a part of a complicated aquatic and benthic biota.

The author has highlighted the role of fish vis-a-vis ornitho food in his books "Wildlife and Ground Flora" and in "Wetland Ecology". Some basic facts on fish are as follows:

Fishes are cold-blooded aquatic vertebrates which breathe through gills and propel and balance themselves by fins.

They are taxonomically divided into :

A. Cyclostomata   : Jawless fishes e.g. lampreys and hagfishes
B. Chondrighthyes : Cartilaginous fishes, e.g. sharks, rays and skates.
C. Osteichtyes   : Bony fishes carp, perch, etc., majority of living fishes.

Fishes as a group has a unique position in water bodies and stream ecosystem. Fish have been always neglected by limnologists despite their:

- Commercial value, being the main produce harvested from inland waters;
- Fish interacts with various levels of the food chains and occupy practically all levels excepting the producer;
- Fish influences the structure of aquatic ecosystems by their highly diversified voracious feeding habits;

CHARACTERISTICS	EUTROPHIC	OLIGOTROPHIC	DYSTROPHIC
1. Depth	Relatively shallow	Very deep	Usually shallow
2. Low water temperature in Hypolimnion	Minimal or Absent	Present	Variable
3. Organic matter in bottom and suspension	Abundant	Very low	Abundant
4. Electrolytes	Variable, Often high	Low or variable	Low
5. Calcium Phosphorus and Nitrogen	Abundant	Relatively poor	Very scanty
6. Humic materials	Slight	Very low or absent	Abundant
7. Dissolved oxygen in deeper water	Minimal or Absent	Height at all depth	Almost absent
8. Larger aquatic plants	Abundant	Scanty	Scanty
9. Plankton : Quantitatively	Abundant	Restricted	Variable (commonly low)
10. Plankton : Qualitatively	Variable	Rich	Poor
11. Water Blooms	Common	Rare	Variable
12. Predominant Phytoplankton	Blue-Green Algae and Diatoms	Green Algae	Blue-Green Algae very rich
13. Profoundal Fauna : Quantitative/Qualitative	Poor in Hypolimnion	Relatively rich	Poor to absent
14. Fauna Type (Benthic)	Chironomus	Tanytarsus	Chironomus (Sometimes)
15. Chaboborus (Corethra) : Midge (Benthic)	Present	Usually Absent	Present
16. Deep-dwelling coldwater fish species	Usually absent	Common to abundant	Absent
17. Fish production	Usually Good	Poor	Very poor
18. Succession into	Swamp or Marsh	Eutrophic type	Peat Bog

TANYTARSUS TYPE : Tanytarsus a midge (insect); its larvae are benthic and intolerant to low O$_2$ levels
CHIRONOMUS TYPE: Chironomus (Tendepes) also a midge; its larvae (Blood water) are benthic and resistant to low O$_2$ levels.

- Fish inhabit almost all habitats from highest elevation to the deepest ocean bottom and over 1.5 km in inland water bodies baikal;
- Fish have acclimatized to temperatures below freezing point at arctic to over 40°C in some hot springs;
- Fish size vary from the minute (12 mm) pigmy goby of the Philippines to the largest whale shark (22.5 m / 25 mt) and giant F.W. Fish of the Amazon river, Arapaima Gigas 5 m / 200 kg/).
- Fish are most numerous [over 20,000 species] among vertebrates and have adapted to diverse ecological and environmental conditions.

Food and Feeding Habitats of Fish

Fishes have highly diverse feeding habits involving all different types of feeding; only one broad characteristic prevails, that is, 'The food is taken into the mouth'. Fish feed on almost all the organisms present in water, such as, zoo and phytoplankton, filamentous, algae, macrophytes, benthic organisms; and also feed on fish themselves. Amphibians, reptiles birds and even mammals feed in fish.

Fish : Major Feeding Types

PREDATORS	: Fishes that feed on macroscopic animals.
CARNIVORES	: That prey on myriad of animal foods from zooplankton and insect larvae to large vertebrates, e.g. sharks.
PISCIVORES	: That mainly feed on other fishes, e.g. seabass, large mouth bass, snake-head (Dalag), grouper (Lapu lapu).
LARVIVORES	: That feed on mosquito and other insect larvae, e.g. guppy.
INSECTIVORES	: Feeding mainly on insects, e.g. some minnows.
GRAZERS	: Fishes that graze and take by bites often by individual small ones. Food is mainly plankton or bottom organisms, e.g. blue gill.
PLANKTIVORES	: That feed mainly on zooplankters, e.g. post-larvae (young fry) of almost all fishes; also bighead, catla.
STRAINERS	: Fishes that feed by straining microscopic organisms mainly phytoplankton, e.g. silver carp.
SUCKERS	: Fishes that suck food into the mouth generally from bottom mud, e.g. catosmids (suckers)

Other Types of Feeding Habits

HERBIVORES	:	Fishes that mainly feed on food of plant origin, e.g. milkfish.
OMNIVORES	:	Fishes that feed on both and animal food, e.g. Tilapia, common carp.
DETRITRIVORES	:	That feed mostly on detritus [decaying leaves, twigs, organic debris, fine silt and mud], e.g. Mrigal.
BOTTOM FEEDERS	:	That feed on bacteria, fungi, protozoans, insect larvae, worms, etc. living on bottom debris, e.g. mud carp.
MACROPHYTOGRAPHS	:	That feed on higher aquatic plants, grass, etc., e.g. grass carp.
MALACOPHAGUS	:	That feed on molluscs (snails, etc.), e.g. black carp.

Forest managers may note the following basic principles relating to Fish in the management of aquatic ecosystem.

- Animals are dependent upon plants for food.
- Fish production in a pond depends on its capacity to produce plants.
- Plant production depends on : sunshine, water, CO_2 and essential elements : C, H, O, N, P, K, S, Ca, Mg, B, Zn, Si and Cu.
- Fertility of watershed limits productivity of natural waters.
- Productivity of water can be increased by fertilization.
- Phosphorus is the first limiting factor.
- Next limiting factor is nitrogen.
- Carbon dioxide is often the next limiting factor.
- Of plants, phyto-plankton appears most desirable for their:
  - short life
  - small size
  - huge nutritive value
  - greater surface area
  - they do not interface with movement of fish
A. Higher aquatic plants (rooted) desirable to limited extent:
  - absorbs nutrients from pond bottom
  - release them into pondwater on their death
  - also form food of fish
B. Floating plants are undesirable because they :
  - cut off sunlight
  - low production of fish-food organisms
  - low oxygen content of water.
- Shorter the food chain higher the production.
- At a certain level fertility, a pond can produce only a certain poundage of fish. The poundage may be a large number of small fish or a small number of large fish.

- Large number of small fish can utilize the available food more fully than a smaller number of large fish.
- For a short period of time the number of fish can be regulated by the stocking rate.
- For longer period of time the number can be regulated by using a piscivorous fish.
- A balance between the population of piscivorous species and forage species to be maintained.
- The greatest weight of a given species can be produced by stocking only that species.
- Maximum production, however, is obtained by using combination of species of different feeding habits.
- It requires more food to grow a pound of fish than to maintain it.

'roductivity and Phytoplankton in Water Bodies

- It is necessary to enumerate some definitions
  - Productivity
  - Standing crop
  - Biomass
  - Yield
  - Primary productivity
  - Production processes
- Measurement of primary productivity
  The calculation processes are not being discussed here.
- Seasonal pattern and periodicity.
- Classification of organism in water.
- Classification and terminology of plankton.
- Functional classification of organism.

Productivity: Rate of production : Amount of accumulated organic matter per unit of time.

Standing Crop: It is the weight of organic material that can be sampled or harvested at any one time from a given area.

Biomass: It is the weight of all living material in a unit area at given time.

Yield: It is the crop expressed as a rate. [kg/ha/yr].

Primary Production: It is the energy accumulated by plants in Photosynthesis.

Primary Productivity: It is the rate of energy storage (per unit of time) by plants or it is defined as the rate at which radiant energy is stored by photosynthetic and chemosynthetic activity of producer organisms [mainly green plants; some Bacteria].

In this connection the following processes are not discussed in this book.

Production Process: Seasonal patterns and periodicity – periodicity of species composition – Classification and terminology of plankton – Functional classification of organism – floating mechanism and water turbulence – Inorganic nutrient factors – Biological factors.

TEMPERATE WATER BODIES

In temperate water bodies, however, there is distinct periodicity in biomass of phytoplankton :

A. WINTER	:	Low light and temperature; growth greatly reduced or negligible; rapid decline in population.
B. SPRING	:	Improved light conditions; increase in phytoplankton numbers and biomass reach maximum; predominant groups; diatoms
C. SUMMER	:	In eutrophic water bodies; brief summer minimum followed by late summer profusion of blue-green algae. In oligotrophic water bodies; low populations.
D. AUTUMN	:	Develop second maximum : Diatoms.

Periodicity of Species Composition: Seasonal community consists of (a) composite or holoplanktonic [perennial species] throughout the year : (b) Meroplanktonic [intermittent special] : Enters some type of Diapause in resting stages.

Winter Populations: Adapted to cold-water and low light intensity; usually small and motile algae – Cryptomonas, Gymnodinium, Dinobryon, Chlamydomonas, Synedra, Fragilaria Trachelomonas.

Spring Populations (Maximum): Mixing of nutrient-laden water; increasing light, dominated by diatoms.

Spring Decline: Reduction in nutrients in Photic zone; decline in diatom population; succeeded by green algae and later by blue-green algae.

Summer Populations: Dominance of Blue-Green algae; Grows ideally at 15°C.

Classification and Terminology of Plankton

i. **On the Basis of Quality :** or [Type of Carbon used as Food]
   Phytoplankton : Plankton of plant origin e.g. Chlorella, Euglena
   Zooplankton : Plankton of animal origin e.g. Brachionus, Cyclops, Moin.

ii. **On the Basis of size :** [=1000 micron (um) or 1 micron (um) = 0.001]
   • Megaplankton : Largest size plankton > 1 cm [Megalo : Very large]
   • Macroplankton : Larger Plankton; visible to naked eye 1 mm – 1 cm [Macro : Big].

- Mesoplankton : Size between macro-micro plankton 50 um – 1 mm [Meso : Medium].
- Microplankton : Small size plankton (Net plankton) 5-50 um [Manno : Dwarf].
- Ultraplankton : Minutest Plankton : observed by electron Microscope < 5 u [Ultra : Beyond].

iii. **On the Basis of Environmental Distribution**
- Limnoplankton : Water bodies plankton
- Rheoplankton : Running-water plankton
- Heleplankton : Pond or Marsh Plankton
- Haliplankton : Salt-water Plankton

iv. **On the Basis of Origin**
- Autogenetic Plankton : Plankton produced locally.
- Allogenetic Plankton : Plankton introduced from other locality
- Tychoplankton : Forms of the littoral community occurring in plankton accidentally

v. **On the Basis of Content**
- Euplankton : True plankton (No debris) [EU: Good, True].
- Pseudoplankton : Debris mingled in plankton [pseudo:False].

vi. **On the Basis of Life-History**
- Holo Plankton : Organisms free-floating throughout their life [Holo : Whole].
- Mero Plankton : Organisms free-floating only at certain times or states of life-history [Mero : Part].

PLEASE NOTE: 'Plankton' is used both for singular and plural. Single organism is known as Plankter.

Functional Classification of Organisms

Organisms can be classified at the functional level by their sources of energy, sources of carbon and in case of lower organisms such as sulphur bacteria by the molecule which serves as the electron donor.

Classification as:

I. Source of Energy
   A. Phototrophs : Organisms which derive energy directly from sunlight through photosynthesis.
   B. Chemotrophs : Organisms which utilize chemical energy as the source.
II. Source of Carbon
   A. Autotrops : Organisms which utilize inorganic CO produce organic matter.
   B. Heterotrophs : Organism which depend on performed organic carbon such as glucose.

III. Source of Electron
    A. Organotrophs : The cells of organisms receive electrons from organic compounds.
    B. Lithotrophs : Organisms that derive electrons from inorganic matter.

Algae are Differentiated as

Auxotrophs : Organisms (algae) that require vitamins for growth.

Photo-Autotrophs : They require light energy for cell metabolism.

Heterotrophs : They sustain growth and cell division in the dark; energy and cell carbon are obtained from metabolism of an organic substance.

Mixotrophs : They assimilate CO_2 in small amounts simultaneously with organic compounds both in light and [especially] in dark.

Floating Mechanism and Water Turbulence

1. Water Turbulence : (Critical in the basic regulation of Algal periodicity and production.)

Importance of water movements in the transport of particulate organic matter is well recognized. These are :
(a) Physical movements of algae into or out of the photic zone.
(b) Vertical transport of mineralized matter from lower depths and Littoral regions to the open water.

2. Floating Mechanisms : (Means to remain within the Photic zone.)

Adaptation to improve floating or reduce sinking rate : Density of most FW Plankton which is 1.01 to 1.03 times of water, causes them to sink **STROKE'S LAW** "Spherical particles of a mean of not more than 0.5 mm Fall."

Various Characteristics to Improve Flotation

(a) Increasing surface-to-volume ratios
    i. Elongation into cylindrical or discoid shapes
    ii. Projection of spines or other cellular processes.
(b) Production of Mucilage [Density of Mucilage is less than the cell]
(c) Gas Vacuoles : These are gas filled structures in protoplasm of living cell
(d) Accumulation of Fat.

Inorganic Nutrient Factors

Importance of phosphorus, Nitrogen and other inorganic macro and micro nutrients for plant growth is well known.

Liebig's Law of Minimum

Each organism requires a certain number of food materials and each of these materials must be present in a certain quantity. If one of these food substances is absent, the organism dies; if not absent but present in minimal quantity, the growth will be minimal. This happens even though ample amounts of other required substances are present.

According to this law, "yield of any organism is determined by the quantity of that particular substance that in relation to the needs of the organism is least abundant in the environment."

Organic matter and Algal growth

Many algae can supplement primary photosynthetic autotrophy by the uptake and utilization of organic substances. They are capable of synthesizing organic growth factors such as vitamins. They are termed Auxotrophic. Some need light energy for cell metabolism [Photo-Autotroph), while others termed mixotrophs which also assimilate CO_2 in small amounts along with organic compounds both in light and especially in dark. A few algae termed Heterotrophs, sustain growth and cell division in the dark; energy and cell carbon are obtained from metabolism of organic substrates.

Biological Factors

Biological factors affecting growth of Phytoplankton are :

(a) Those which compete for essential available resources
(b) Micro- or macro-organisms predating on algae.

Concluding Remarks

Each species of algae possesses a range of tolerances to extreme of these factors and growth proceeds best at a given optimal combination of the interacting factors.

Parasitism and Grazing

A. PARASITISM   : Parasitism in phytoplankton algae has been observed in recent years. Best known parasites are biflagellated phycomycetous fungi; also bacterial and viral infections of algae occur.

B. Grazing by Animals : Predation on phytoplankton by zooplankters, particularly by rotiers and micro-crustaceans is a significant factor in the decline of algae.

Cyanophyta or Myxophyta

[Blue-Green Algae]

- Photosynthetic, prokaryotic microorganisms; some capable of N_2 fixation.
- Mitochondria, Golgi apparatus, plastids and chloroplasts absent; also lack membrane-bounded DNA.
- Spores resistant to adverse conditions.
- Heterocysts : Thick-walled, translucent; formed of vegetative cells – site of N_2 fixation.
- Unicellular, colonial or filamentous forms having gelatinous sheath.
- Reproduction primarily by binary fission.

Chlorophyta

[Green Algae, Desmids and Filamentous forms]

General Characters

- Unicellular, colonial or multicellular forms.
- Chlorophyll a and b
- Cell-wall of cellulose or pectin
- Reserve food-starch

Examples : Chlorella, scenedesmus, desmids and fil. algae.

Desmids

Cell-wall in two sections which adjoin at the mid-region forming two "semi-cells" with the chloroplast correspondingly divided.
Examples: Cosmariu, closterium.

Filamentous Green Algae

Branched or unbranched multicellular forms.
Example : Spirogyra

Euglenophyta [Euglenoids]

- Unicellular with cell-walls.
- Capable of Photosynthesis (chlorophyll a and b, chloroplast present).
- Reserve food : Paramylum bodies (starch-line carbohydrate).
- Flagella present; usually two (one long, one short).

Charophyta [Stonewort]

Erect, branched thallus differentiated into a regular succession of nodes and internodes.

Bacillariophyta [Diatoms]

- Golden-brown in colour (carotenoid pigments).
- Unicellular, sometimes adhere to one another forming chains.
- Cell-wall of silica with minute clear markings, composed of two valves.
- Each diatom cell contains one nucleus.
- Reproduction usually vegetative; sexual by auxospore formation
- Important group of phytoplankton

Two groups of diatoms :

(a) Centric Type : Exhibits radial symmetry, e.g. Cyclotella.
(b) Pinnate Type : Exhibits bilateral symmetry, e.g. Navicula, Pinnularia.

Pyrrophyta and Rhodophyta

Pyrophyta (Dinoflagellates)

- Unicellular : Flagellated; motile
- Cell-wall absent or if present – composed of cellulose
- Chlorophylls a and c
- Food stored as starch
- Mostly marine
  Examples : Peridinium, gymnodinium, gonyalaux, ceratium, etc.

Rhodophyta

- Colour red or dull purplish green.
- Mostly marine forms : Order : Florideae freshwater.
- Multicellular thalli; cell-wall contains cellulose and protein.
- Chlorophyll a and b; blue (phycocyanin) and red (phycoerythrin) pigments.
- Reserve food: Floridean starch.

Chrysophyta

A. Chrysophycease [Golden-Brown Algae]

- Chromatophore: Golden-brown colour due to β-carotene and Xanthophyll carotenoids; have chlorophyll a.
- Unicellular or colonial; rarely filamentous.
- Flagella : one or two of equal length.
- Many without cell-wall; only bounded by cytoplasmic membrane.
- Vegetative reproduction by longitudinal cell division is most common.

Examples : Crysococcus, Binobryon, Sinura.

Xanthophyceae [Yellow-green algae]

- Yellowish-green chromatophores; chlorophyll a and c, β-carotene and Xanthophyll.
- Unicellular, colonial or filamentous. Two flagella one longer than the other.
- Cell-wall is often absent. Starch is never formed.
- Reproduces mainly asexually by aplanospores.
  Examples : Chlorobotrys, gloechloris.

Cryptophyta

Cryptomonads : [Cryptophyceae]

- Naked, unicellular and motile; two flagella equal in size.
- Chlorophyll a, c, d and e, carotene, biliproteins present
- Stored food : starch
- Reproduction by longitudinal division.
  Examples : Cryptomonas, Rhodomonas, Chroomons

Zooplankton and Zoobenthos

Phyla Protozoa, Rotifera, Arthopoda, Cladocera and Copepoda have innumerable species form the population of zooplankton in water. Some information on their morphology and common species are presented in short.

A broad based information on the occurrence of zoobenthic fauna is also presented. These animals are of vital importance to fish and quake birds.

Phylum : Protozoa

Protozoa	:	Microscopic, single-celled or in colonies of similar cells; no tissues protozoans feed on detritus, bacteria, fungi yeasts and other protozoans.
Subphylum : Plasmodroma	:	Organelles for locomotion are flagella, pseudopodia or none; nuclei of one kind.
Class : Sarcodina	:	Pseudopodia for locomotion, e.g. amoeba, difflugia, atmospheres, arcella.
Subphylum : Ciliophora	:	Cilia or sucking tentacles in at least one stage of life-history; nuclei of 2 kinds
Class : Ciliata	:	Having cilia for locomotion, e.g. Coleps, stentor, paramecium, vorticella, lacrimaria, stylonychia.

Phylum : Rotifera [Rotatoria or Wheel Animalcules]

Characters

Microscopic forms (40 μm - 3 mm); Body distinguishable to head, trunk and foot.

A ciliated corona in the head; mouth located in corona (wheel). Trunk covered by cuticle; in some thickened to form lorica (rigid).

Foot telescopically retractable; bears 1-4 toes. Pharynx muscular [mastax] having hard jaws (trophi) for sucking and grinding food particles.

Feeding Habits

Detritus feeders : Feeding by means of ciliary movement of corona directing food to mouth e.g. Brachionus, Keratella, Lecane

Carnivores (by grasping and trapping) : Trophi used as forceps; Prey seized, Macerated, Edible parts sucked in and rest discarded e.g. Asplanchna.

Phylum Rotifera [Rotatria or wheel animalcules]

Characteristics

- Microscopic forms (40 μm - 3 mm); body distinguishable to head, trunk and foot.

- Head bears a ciliated corona (resembles wheel); mouth located in corona.

- Trunk covered with a transparent cuticle; in some thickened to form Lorica (rigid).

- Foot telescopically retractable; bears 1-4 toes.

- Syncytial epidermis and pseudocoil present.

- Digestive system complete; pharynx muscular (mastax), equipped with hard jaws (trophi) for sucking in and grinding food particles.

- Salivary and gastric glands present

- Excretory system with a pair of protonephrial tubules with flame cells.

- Nervous system with bilobed brain and sense organs.

- Reproductive system with combined ovaries and yolk glands (germovitellaria); parthenogenetic females produce diploid eggs hatching to diploid females.

Feeding Habits

Detritus feeders : Feeding by means of ciliary movement of corona directing food to mouth cavity; e.g. Brachionus, Keratella, Lecane.

Carnivores (by grasping and trapping) : Trophi used as forceps; prey seized macerated, Edible parts sucked in and rest discarded,e.g. Asplanchna.

Phylum Arthropoda [Animals having jointed legs]

Class Crustacea	:	Hard shells; two pairs of antennae present.
Subclass : Branhiopoda	:	Legs are flattened and leaf-like with gills the chief respiratory organs.
Order : Anostraca	:	No chrapace, e.g. fairy shrimps and brine shrimps.
Order : Notostraca	:	Carapacelow, e.g. tadpole shrimps.
Order : Conchostraca	:	Carapace bivalved, e.g. clam shrimps.
Order : Cladocera	:	Minute, usually bivalved carapace, e.g. Water fleas.
Subclass : Ostracoda	:	Minute, bivalved, compressed and enclosing entire body, e.g. cypris, cypria.
Subclass : Copepoda	:	Mostly microscopic; 9 free trunk somites; no appendages in last four somites.
Order : Calanoida	:	e.g. Calanus (exclusively marine), diaptomus.
Order : Harpacticoida	:	e.g. Harpacticus
Order : Cyclopoida	:	e.g. Cyclops
		Produce eggs which develop into parthenogenetic and without fertilization. Number of eggs produced per clutch varies : 2 (Chydoridae to 40 (Daphnidae); Eggs are deposited in a broad chamber or pouch; eggs develop in the pouch and hatch in a small form similar to the present.
Sexual Reproduction	:	Along with Rotifer.

Copepoda

Characteristics of free living Copepoda

Body consists of 1. Metasome (Cepwalotnorax); head with 5 pairs of appendages of antennae and mouth parts; thorax with 6 pairs of swimming legs.

2. Some abdominal segments (in 1st segment modified as genital segment) and terminal candal rami bearing setae.

Feeding

Calanoids feed chiefly by filtration of plankton, the antennae being used to produce a current.

*Cyclopoids have the mouth parts modified for seizing and biting. Food consists of unicellular plants and animals and also organic debris.

Harpacticoid mouth parts are adapted for raking, seizing and scraping food from the bottom.

*Carnivores	:	Mesocyclops, Cyclops
Herbivores	:	Eucyclops, Acanthocyclops

Zoobenthos

Major Zones in Waterbodies Floor

1. Littoral } zones of diversity}	Zoobenthos
2. Sublittoral} and high productivity}	Most types of insects, snails, worms, crustacean and fish found. Aquatic plants are abundant in shallow zone.
3. Profoundal zone :	Below thermocline, uniform physico-chemical factors except decrease in O_2 in cummer (Eutrophic Waterbodies).

Zoobenthos : Four Main Groups

1. Chironomid midge larvae, 2. Oligochaete worms, 3. Chaoborus (Phan Midge larvae), and 4. Fingernail clams.

Seasonal changes (expected) in Benthos of an average temperate Waterbodies of the second order.

SPRING

- Rapid growth of plants.
- Periodic insect emergencies.
- Upward migration of snails (To breed).
- Increase in population of various benthic invertebrates.
- Beginning stages of summer concentration zone.
- Migration of Sialis larvae to shore.

SUMMER

- Culmination of plant growth.
- Fluctuations of population by insect emergences.
- Shoreward migration by certain mayfly nymphs.
- Maximum for various chironomid larvae.
- Continuance for growth and reproduction in various invertebrate.

WINTER

- Minimum of plant growth.
- No insect emergences.
- Increase in population due to development of autumn-laid insect eggs.
- Reduction in reproduction, growth and other physiological processes.
- Midwinter maximum tubificidae and vhaoborus larvae.
- Gradual reduction of benthos to late winter minimum.

CLASSIFICATION OF BENTHIC INVERTEBRATES COMMON IN WATERBODIES AND STREAMS

Taxonomic (Flatworms)	Food/Feeding Habit	Examples
Torbellaria (Flatworms)	Carnivores	Dugesia
Nematoda (Round works)	Carnivores, Herbivores	Dolichodorus
Annelida Oligochaeta (Earthworms)	Sediment grazers	Tubifex (Red worms)
Hirundinea (Leeches)	Carnivores, Detrivores	Haemopsis (Horse Leech)
Mollusca Gastropoda (Snails)	Grazer	Pila, Limnaea (Pond Snails)
Pelecypoda (Bivalves)	Filter Feeder	Planorbis
		Anodonta (Swan musel)
Crustacea Malacostraca		
(Crayfish Amphipods)	Detrivores	Cammarus, Astacus
Insecta Megaloptera (Dobson-Flies)	Carnivores	Sialis
Pleooptera (Stone Files)	Mostly Carnivores	Nerouba
Odonata (Dragon Files)	Raptorial, Carnivores	Lebellula, Anax
Ephemeroptera (May Flies)	Mostly scrapers, Grazers	Baetis, Ephemerella
Hemiptera (Water bugs)	Breaked Carnivores, Herbivores	Notonecta (Backswimmers)
Trichoptera (Caddis Flies)	Mostly filter feeders Scrapers	Limnephilus, Hydropsyche
Coleoptera (Beetles)	Raptorial carnivores	Dytiscus, Cybister
		(Diving beetles)
Diptera (Two winged Flies)	Filter Feeders	Culex (Mosquito)
Family : Culicidae	Raptorial carnivores	Chaoborus (Corethra)
		(Phantom Midge)
Family Simulidas	Filter Feeder	Simulus (Black Fly)
Family Tendipedidas (Chironomuds)	Filter Feeder	Tendipes (Chironomus)
		(Blood worms, True Midge)

Vertical Distribution of Zooplankters

Since conditions change with increasing depth the vertical distribution of plankers is quite complicated. Generalization is difficult.

Tendencies

(a) Greater occurrence of sarcodina in lower waters.
(b) Preference of dinoflagellata for upper waters.
(c) General scattering of the ciliates.
(d) Selection of different levels of young and adult stages of certain crustacea and utaers.

Vertical distribution is affected by various events of stratification. During periods of thermal and chemical stratification, certain plankters form distinct concentration of population at some particular depth.

Conditions Influencing Vertical Distribution

1. Light, 2. food, 3. dissolved gases and other substances, 4. temperature, 5. wind, 6. gravity, and 7. age of individuals.

Vertical Migration of Cladocera and Copepods

More conspicuous in Claodocera than Copepods, light is the controlling factor for V.M. but geotactic reaction may also be involved.

Nocturnal Migration (Upward movement at evening)

Most species migrate upward from deeper waters to more superficial strata as darkness approaches. Single maximum occurs in between sunset and sunrise. Vertical convection current also helps V.M.

Due to

- Predation pressure.
- Variable food quality of algae; protein synthesis is maximum during night.
- Growth efficiency is greater at low temperature.

Twilight Migration 2 maxima occur in surface layers at dawn and dusk.

Reverse Migration (Downward movement due to negative phototaxis).

Single surface maximum during daylight period. More common under ice-cover (Copepoda).

Molluscs in Waterbodies

The role of mollusc in waterbodies of India is of much importance and the forest manager of protected areas must have at least a broad knowledge on the subject.

A short note on (i) molluscs of commercial importance; and (ii) morphology and classification is presented which will be very useful for the management staff of protected areas (aquatic) is presented.

Most of the species, provide edible recipe for wild animals and man. Several fresh and estuarine – marine species are widely cultivated.

Molluscs' Commercial Importance

Clam : Cockle : Scallop : Mussel : Oyster : Abalone

Clam : Clams are bivalves with long siphons, equal shells closed by two adductor muscles and a powerful burrowing foot – both marine and fw forms – 20,000 edible species – 50 commercial – most inhabit shallow waters – size ranges from 0.1 mm to 1.2 m – usually remains buried from just beneath the surface of depths about 0.6 m – FW clams : Glochidium larvae (parasitic) – marine clams : Free swimming larvae : veliger – example : Giant clam Tridacna gigas, Solidissima mercenaria mercenaria

Cockles : Cockles are marine bivalves of the family cardiidae – 250 species – cockle has 2 heart- shaped shell of equal size and shape with serrated edges of many radial ribs – has a strong muscular foot can move easily size 1.2 cm to 15 cm – most species live just below low tide line, some in depth up to 3 cm. Example: Giant pacific cockle Trachycardium quardragenarium, great heart cockle Dinocardium robustum.

Scallop : Fall shell or comb shell, marine bivalves belonging to family pectinidae – 50 genera 400 species habitat – intertidal zone to considerable depths – size 5 cm. Single adductor muscle can swim rapidly by forcibly ejecting water edges of mantle have many well-developed ocelli or eyes – larvae free swimming example : Giant deep-sea scallop - 1 plecopecten maggellanicus : Atlantic bay scallop aquipecten irredians.

Mussels : Mostly comprise fresh water bivalves of the families unionidae (750 species) margaritanidae and mutilidae – marine bivalves mostly belong to family mytilidae – distribution nearly all temperate

		and tropical seas of the world clinging to rocks in tidal zones attached bi valves are elongated, pointed one end, e.g. Mytilus edulis (brown mussel) perna viridis (green mussel).
Oyster	:	Bivalves of family ostreidae (70 spp. Out of 400 spp. Commercially important) inhabits temperate and tropic seas – two shells unequal; left is bigger; right is thin and flat and less developed – sedentary type is small entirely lacking in some-single large adductor muscle.
Abalone	:	Large marine snails belonging to the family species only about 20 commercial species there are holes on the left for respiration single coiled shell. Example : Haliotis discus.

Mollusc : Phylum Mollusca (Mollis = Soft)

- Soft unsegmented bodies : usually bilaterally symmetrical
- Body surrounded by thin fleshy mantle
- Commonly sheltered in an external llimy shell secreted by mantle
- Many with prominent ventral muscular foot
- Mouth with radula unique rasping organ (except bivalves)

Class : Ambineura

Foot flat, shell with series of flat plates (Amphineura has 3 orders which are now classified as classes).

- Aplacophora : Worm like solenogastre (Neomenia).
- Monoplacophora : e.g. Neopilina.
- Polyplacophora : e.g. Chiton.

Class : Scaphopoda

- A small group with conical shell.
- Open at both ends : dentalium (Tooth shell or tusk shell).

Class : Pelecypoda (Bivalvia) (The largest commerically important Molluscan group)

- Foot spade or axe shaped.
- Foot lateral shells hinged dorsally e.g. Clam, Cockle, Scallop, Mussel, Oyster, etc.

Class : Gastropoda (Bilateral symmetry is frequently obscured by a secondary change of torsional nature).

- Broad ventral flat foot.
- Shell single usually coiled e.g. snail, slug, abalone, etc.

Class : Cephalopoda

- Body divided into a large head and sac like trunk.
- Foot surrounds mouth forming tentacles / siphons.
- Fastest moving and most developed molluscs, e.g. Octopus, squid, nautilus, etc.

How to distinguish Mullusc groups?

Mussel, clam, cockle, scallop, oyster, abalone

Mussel :
- Anterior end pointed – post end broader – attaches byssus threads.

Clam :
- Clams have long siphons
- Shells of equal size
- Burrowing habit but moves with powerful foot, e.g. Meretrix meretrix (hard clam)

Cockle :
- Heart- shaped
- Burrowing
- Sedentary
- Can move easily, e.g. Anadora suberenata

Scallop :
- Fan or comb-shaped
- Moves fast throwing water on two sides e.g. Pecten yessoensis

Oyster :
- Sedentary unequal shells
- Left shell (lower) larger than the right (upper)
- Single adductor muscle
- Foot small, e.g. Crassostrea gigas

Abalone :
- A gastropod having single shell
- Holes along the left side of the shell, e.g. Haliota sp.

10

History of Forest Administration and Conservation in India
— Evolution of Forestry in India and Conservation and Development of Resources

*T*he present treatise of 300 pages can hardly relate the elaborate history of forest conservation and administration as it will need about thousand of pages even to draw an outline of the history of forests and foresters. The readers must not expect various issues that the foresters are concerned with today and the activities performed by them in this short presentation.

Forestry is one of the noblest professions on earth and foresters. "The Angels", are meant to perform to create a better environment, for playing constructive role and protect humanity from destruction of forest.

Three brilliant officers who shaped Indian Forestry in various fields are Sir (Dr.) D. Brandis, Dr. W. Schleich and Dr. B. Ribbentrop.

They had vision and profound knowledge on a wide variety of subjects. It is because of foolproof recruitment systems introduced by Dr. Brandis, the father of Indian Forestry, that the forest service could create a tradition of a large group of sincere and efficient officers of strong character. Dr. Brandis felt that a forester should have tact, a sense of sound judgement, high moral character, good family linkage, even temper, good health, super abilities and humane in dealing with local people.

For this rigid recruitment procedures the forest service personnel all over the India had a distinct tradition and identity other than disciplined army and police personnel. This tradition has been maintained over the last 200 years.

However, the people in general and the intelligentsia in particular know little about forest personnel and the nature of their duties and that they have to perform multifaceted duties. The foresters administer 20% of the country's land with bare minimum resources and amenities.

In this chapter various facets of forestry and role forests have been brought to light in a nutshell.

History of Forest Administration and Conservation In India

— Evolution of Forestry in India and Conservation and Development of Resources

An eventful and fact-based history of about 200 years can't be brought to light in a treatise covering 300 pages. Various facts of forest and forestry covering the vast country India as it was in early nineteenth century and thereafter may need thousands of pages or more to describe even in nutshell. As such constructive efforts and salient achievements in colonial period have been brought to light in this chapter which will only be a glimpse of the activities of foresters. So this presentation will not give the complete picture of foresters' effort and achievements.

India does not have any recorded history of its forests. The Ramayana and Mahabharata mention about forested tracts of Chitrakut, Khandara Nana, Dandakaranya and Panchavati. The *Vedas* and *Upanishads* mention about the forest and wild animals, but those do not give any chronological history of forests.

Shikar (hunting) episodes of Mughal emperors show the occurrence of dense forests near Delhi and the presence of elephants, rhinoceroses and other wild animals near Delhi. During Alexander the Great's invasion some historians mentioned about dense forest near Taxilla, though evidence of dense forests may be found in the burning of Khundava forests. As such it is obvious that there were no systematic account of forests, it's men and activities concerned it in any of the ancient literatures.

From the beginning of Colonial rules in India multifaceted information on socioeconomic, cultural, economic, natural history, flora, fauna, etc. have been recorded in administrative reports, district gazetteers, census reports, revenue and various other documents. Authentic literatures exclusively on forestry administration and conservation are available some of which are:

- *The Forest of India* by E.P. Stebbing (Vol. I, II, III and IV).
- *Forestry in British India* by B. Ribbentrop.
- *Manual of Indian Forestry* by Sir W. H. Schleich.

and many other administrative reports which were master pieces of creation of colonial foresters and administrators.

In the post independence period two volumes during centenary (100 years) of Indian Forestry were published by Govt. of India which gives briefly the history of the years from 1864 to 1964. The Forest Department, Govt. of West Bengal also published its forestry centenary issue entitled *100 Years of Forestry in West Bengal*.

Ribbentrop was a visionary as Brandis and Schleich were. He visualised the deterioration of climatic conditions due to destruction of forests. He particularly mentioned about the denudation of forests in dry zones which had once thick cover of forests. The damage caused by fire, grazing and shifting cultivation caught his eyes (Chapter 3).

Although the colonial administrators pinpointed teak (*Tectona grandis*) as a substitute of British Oak and initiated actions to regulate wanton felling of the species and started creation of plantations. Real forestry started sometime in the late 1850s after Lord Dalhousie's memorandum – *Charter of Forestry* (Aug. 3, 1855) outlined the policy declared the Forest Conservation and administration in India.

Some Facts on Works Initiated in Early Years

- Scientific knowledge of the European officials was confined entirely to the members of medical profession.
- First plantation was created in Malabar coast between 1805-1827; then in Burma, Bengal and elsewhere.
- In 1806 (November 10th) Captain Watson of Police was appointed as the first Conservator of Forests.
- Tipu Sultan, the ruler, enforced restriction over the felling of Teak as early as in 1792.
- For sustained timber production the British officers did not have the expertise; so they sought the help of a number of competent German foresters.
- As early in 1888 Dr. Ribbentrop mentioned about the availability of 1200 tree species in India (some of the sought after species were Teak, Sal, Deodar, Sissoo, Setisal, Klair, Sandal, Babul, Toon, Red Sandaurs, etc.
- The British foresters undertook the planting of Teak, Eucalyptus, Casuarina, Sissoo, Acacia in many parts of India since mid 1830.
- The collector of Malabar, Conolly planted teak in Malabar in 1840 considering the species as a substitute for the British Oak.
- Construction of Forest Research Institute, Dehradun was complete in 1926.

Forest Policy of India

It went through three phases in post independence period.

1952–76	– Focus was for timber
1976–88	– Focus was on –
	– Commercial forestry on forest land
	– Social and farm forest in non-forest and private land
1998 onwards	– Joint Forest Management and integrated forestry.

1985-90 i.e. 7th Plan document for the first time recognised the importance of non-market and ecological benefits from forests. It did not explicitly maintain producing timber for commercial purposes as one of the object of forest policy. After meeting the needs of the local people raw materials could be provided to forest based industries.

In 1987 at the Central Board of Forests meeting it was decided that Forest laws would be used for preserving soil and water system, and not for generating state income. All supplies to match and other industries would be met from farm forestry. In May 1988 the forest minister announced in the Parliament that 70 percent of the total afforestation would be in the farm sector. The Central Board of Forestry also took a courageous step in recommending a ban on commercial exploitation of degraded forests, and on replacement of natural forests by monocultures. The policy gave higher priority to environmental stability than to earning revenue.

The policy gave importance and priority to:

- Tribal rights of forest produce.
- Minor forest produce.

In June 1990 guidelines for the implementation of the policy was facilitated by the Govt. of India to make it possible to involve people in the management of forests.

The Forest conservation Act, 1980 limits the powers of state governments, and they cannot de-reserve forests or divert forest lands for non-forest purpose without permission from Govt. of India.

Various anthropogenic factors – unemployment, illiteracy, lack of accurate data and absence of strong political will to fight all odds refrained the policy makers in framing a foolproof forest policy.

Tens of millions of hectares of barren areas badly need to be brought under vegetal cover, yet degradation and depletion of forest cover continue in an alarming scale. In addition forestry sector has too inadequate budget allotment to meet all ends.

Forestry has to provide improved economic status to the tribal people, labour force. Simultaneously the amenities of the foresters are to be improved and salary enhanced as they have to work in areas, devoid of urban amenities.

It is necessary that the forest managers have specialised knowledge in specific fields such as forest botany, ecology, conservation, genetics and tree-breeding, plant physiology, plant pathology, entomology and wildlife besides many other relevant fields.

Forestry research has to be geared up manifold. Forestry extension services have to cover the curricula of schools and colleges.

Forestry services have had always highly qualified personnel having M.S. and PhD degrees, who have to be given extra facilities to devote more time for research work in each of their specialised fields. Moreover all kinds of forestry operations need to be modernised.

The authors feel that the training of forest officers and staff should be as rigorous as it was during colonial period.

The major reasons for depletion of forest resources have been due to:

- inability to preserve tree cover,
- inadequate technical staff and modern equipment,
- inadequate allocation of financial and managerial resources for afforestation,
- failure to bring more and more barren areas under afforestation,
- pressure of a vast poor population for fuel, fodder and other materials,
- failure to intimately involve fringe and tribal people in conservation and augmentation of resources,

In recent years foresters have made approaches on 'Eco-Management Plan', 'Microplan' and 'Management of forests for intensive forestry works, involving fringe population'.

Forest Rangers

They are the backbone of Forest Administration and Management

Sir Harry Champion commented about the training of the Forest Rangers as follows :

"..... the success of the Forest Department has been, and is closely related with the success in the building up of Forest Ranger cadre. As Range Officer, a Forest Ranger acts as local deputy to the D.F.O. responsible for running the range. The D.F.O. has to decide as to what has to be done to carry out working plan prescriptions and directions of higher administration, but it is the F.R. (R.O.) who has to do it or get it done. No. D.F.O. can run division well unless he has competent and reliable Rangers."

FOREST RANGERS ARE THE BACKBONE OF FOREST ORGANIZATION

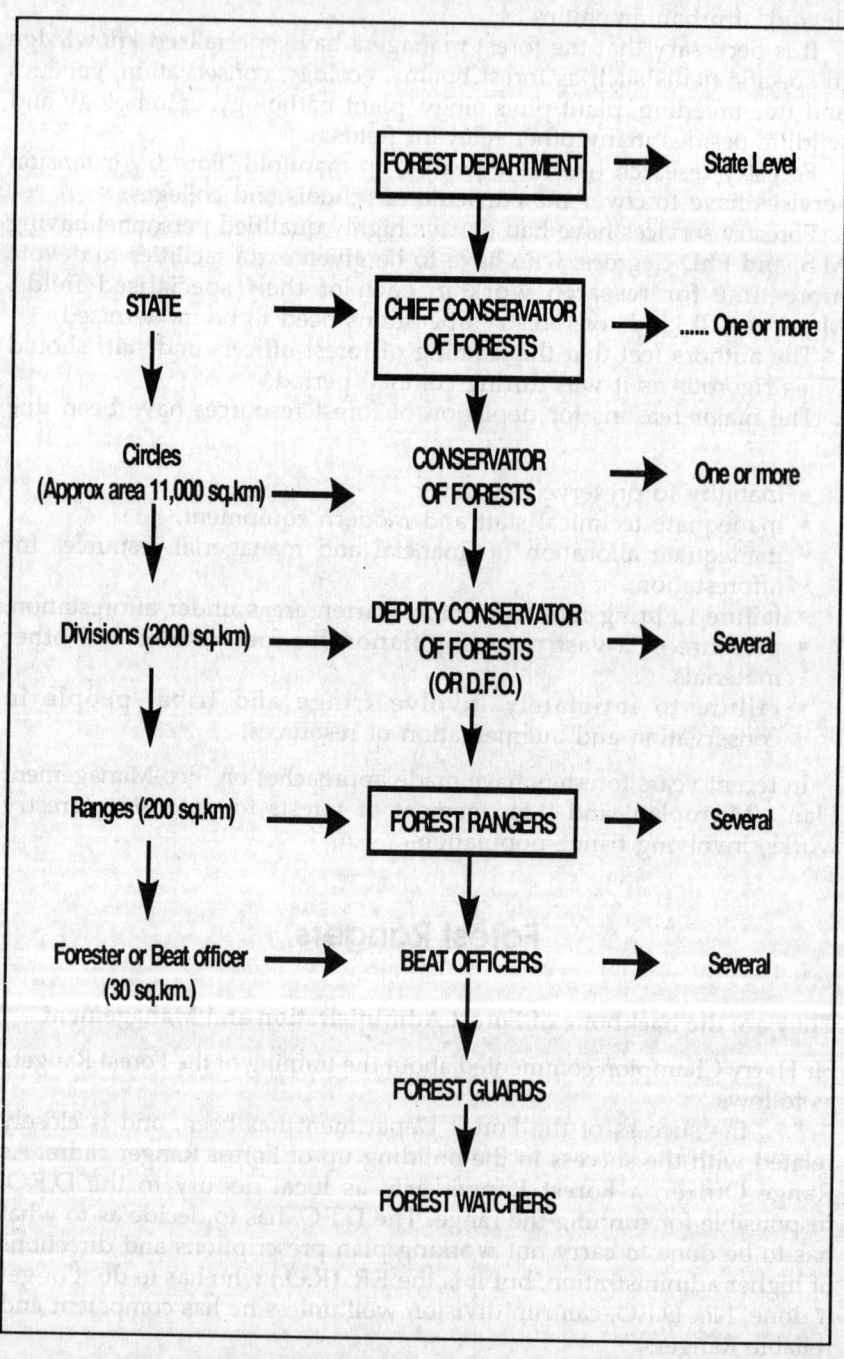

Duties of Forest Ranger

- He is the head of his Range jurisdiction.
- All the activities of entire area is executed under his direct command and supervision.
- He prosecutes persons, involved in forest offence.
- He checks boundary demarcation, through boundary survey.
- He maintains regular liaison with D.F.O. for day-to-day works.
- Looks after the maintenance of departmental elephants.
- He prepares marking lists for sale of timber.
- He realises revenue on sale of forest produce and deposits to the Treasury.
- He prepares regeneration plan and working scheme for raising plantations.
- He keeps the D.F.O., well-informed about the latest condition of his jurisdiction.
- He maintains plantation registers and maintain the same making it up-to-date.
- He performs the job of a civil engineer in constructing buildings roads and bridges and maintenance of the same.
- He performs the jobs of a police officer in apprehending and effecting arrests of offenders prosecuting them in the court of law and empowered to act as a judge in assessing of compensation and compounding of cases and set the offender free. He can also release offender on bond, subject to prosecution in the court of law; he also prepares seizure list of produces seized and submit the seizure report to S.D.J.M., in connection with Prosecution Offence Report (POR).
- He maintains cash book for the entire transactions against different works including payment of salaries and other allowances himself and the staff under his disposal.
- He maintains permanent store register for movable Govt. properties. He also maintains the running register for issue and return of different stores to and from the staff.
- He attends the inspection of his superiors and submits reports of compliance of inspection reports.
- He maintains order-book and submits copy of the same with the remarks of superior officers with notes of compliance on progress.
- He attends the inspection of superior officers and ensures the safety and easy-stay of him during the period of his stay.

In fact the Forest Ranger is the master of the entire activities of his Range jurisdiction. He is an officer bestowed with multiple qualities of tenacity, patience, tact, dedication, discipline and devotion.

Conservation Tools

The colonial foresters had the qualities of top class administrators, who had visionary and sincere approach in the establishment of forestry in India. They never thought of exploitation of forests only to enrich their exchequer; instead, they planned the management for the benefit of local people on a sustained basis.

They realised that vast forest resources were being mutilated in various ways and to save these resources a sound policy, Acts and Rules were necessary. The chart presented hereinafter records some of the efforts made by erstwhile foresters that will prove how visionary and sincere they had been and worked for the benefit of people.

CONSERVATION TOOLS APPLIED BY FORESTERS

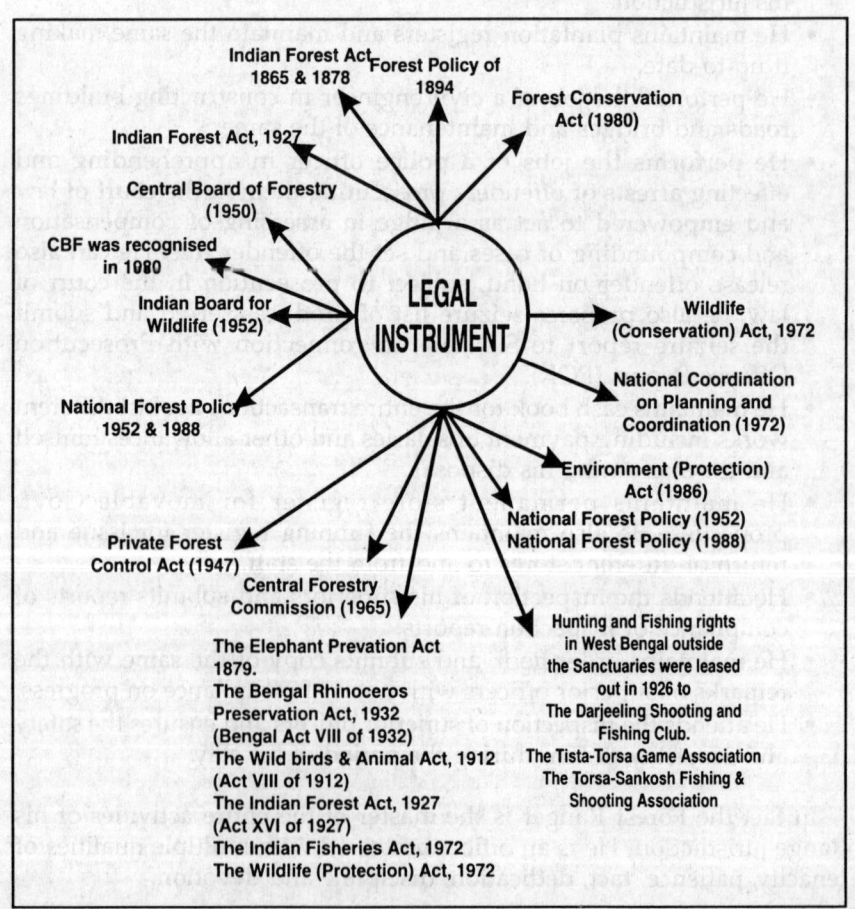

Indian Forest Act 1865 & 1878

Forest Policy of 1894

Forest Conservation Act (1980)

Indian Forest Act, 1927

Central Board of Forestry (1950)

CBF was recognised in 1980

Indian Board for Wildlife (1952)

National Forest Policy 1952 & 1988

LEGAL INSTRUMENT

Wildlife (Conservation) Act, 1972

National Coordination on Planning and Coordination (1972)

Environment (Protection) Act (1986)

National Forest Policy (1952)
National Forest Policy (1988)

Private Forest Control Act (1947)

Central Forestry Commission (1965)

The Elephant Prevation Act 1879.
The Bengal Rhinoceros Preservation Act, 1932 (Bengal Act VIII of 1932)
The Wild birds & Animal Act, 1912 (Act VIII of 1912)
The Indian Forest Act, 1927 (Act XVI of 1927)
The Indian Fisheries Act, 1972
The Wildlife (Protection) Act, 1972

Hunting and Fishing rights in West Bengal outside the Sanctuaries were leased out in 1926 to
The Darjeeling Shooting and Fishing Club.
The Tista-Torsa Game Association
The Torsa-Sankosh Fishing & Shooting Association

Background Policy of 1894

After the demarcation, mapping of Forest areas the Govt. of India invited Dr. Voeleker to examine the State of agriculture of the country. Voelker's report (1893) discussed at length the condition of the forests and stressed the need of formulating a forest policy with a definite bias serving agricultural interests more directly. On the basis of his recommendation, the Govt. of India issued Resolution No. 22 F dated. 19th October, 1894 declaring its forest policy.

The basic principle of this policy were:

- The state forests are to be administered in public benefit.
- The regulation of rights and the restriction of privileges of the user of the forests.
- Hill forests are to be maintained as protection forests to protect cultivation in plains.
- Valuable forests are to be managed on commercial line.
- Honeycombing of valuable forests for cultivation was not allowed.
- Inferior forests are to be managed in the interest of the local people.

Indian Forest Act of 1878 succeeded Forest Act of 1865; and preceded the Indian Forest Act, 1927.

A perusal of the charts will reveal how seriously the colonial foresters thought of formulating Acts to protect various resources. In post independence period the Indian foresters have also taken adequate steps which proves their sincere conservation efforts.

Some of the objectives of National Forest Policy (1952 and 1988) have been shown in the following charts : (This chart is onto the next page)

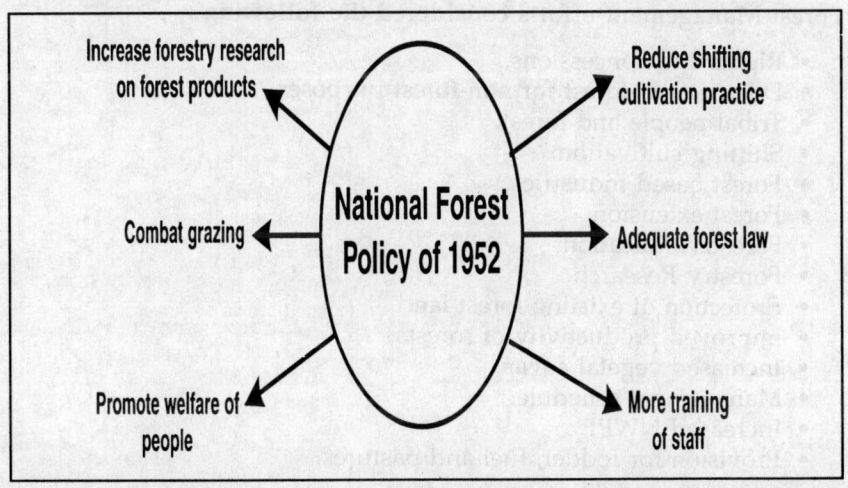

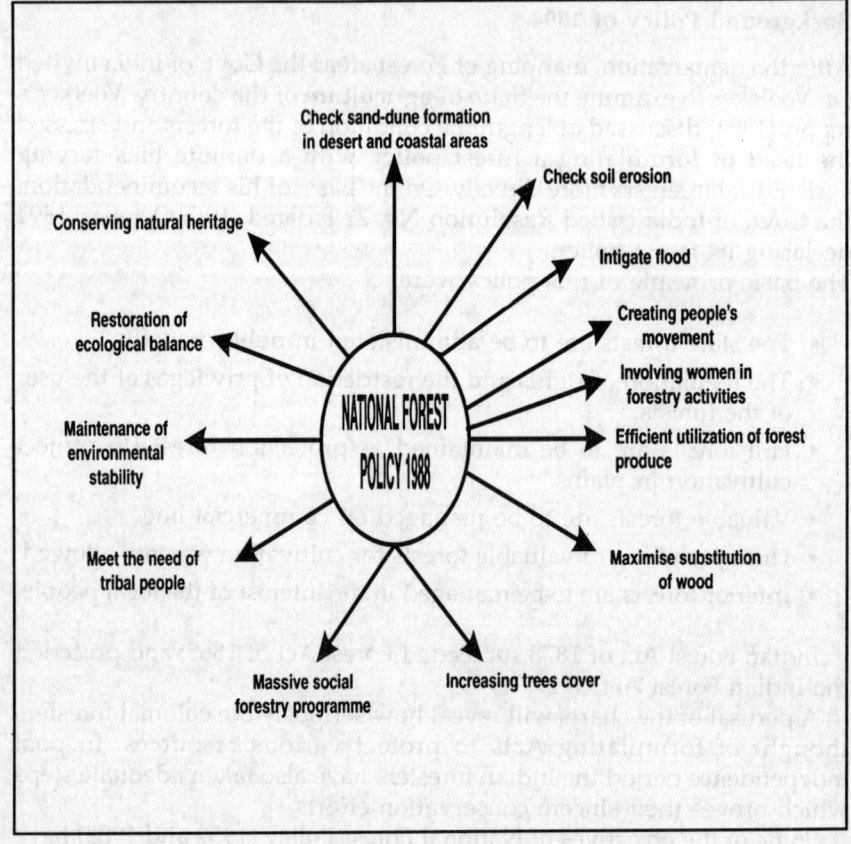

Forest Management efforts considered the following :

- Rights and concessions.
- Diversion of forest for non-forest purposes.
- Tribal people and forest.
- Shifting cultivation.
- Forest based industries.
- Forest extension.
- Forestry education.
- Forestry Research.
- Protection of existing forest land.
- improved productivity of forests.
- Increased vegetal cover.
- Management schedule.
- Increased NWPF.
- Provision for fodder, fuel and pastures.

PLANTATION FORESTRY: Forester's unique and invaluable contribution to enhance resources

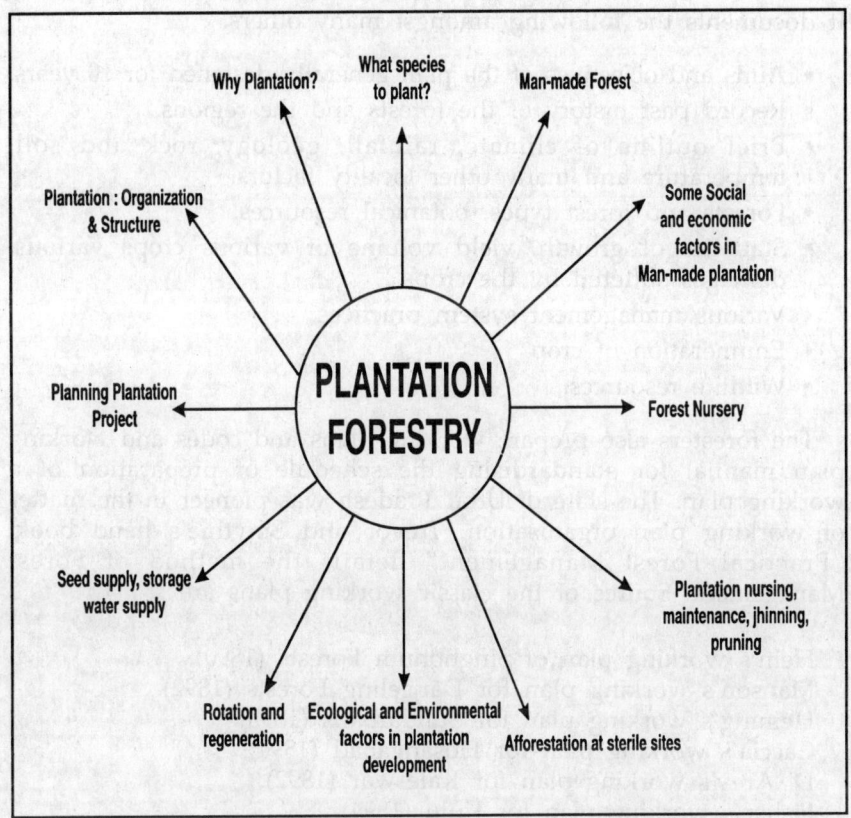

Working Plan Is A Management Tool

It is a forester's endeavour, forethought with advance planning as to how to work for creation for; (i) sustainable yield, (ii) to satisfy local people's need for timber fuel wood and minor forest produce, etc. (iii) creating timber and wildlife resources, (iv) protecting forests and several other such functions. In managing agricultural crop one year's planning may be enough, but in forestry a crop planted to day matures after many decades and a plan has to be written based on various growth and very many other factors.

Other aims of a working plan are to protect physical feature, ameliorate climate, improve water supply and meet the requirement of local people.

The concept and mutifaceted approach in preparation of a working plan document show the foresight and visionary qualities of the then foresters. It is a masterpiece of forester's creative genius. It documents the following amongst many others.

- Aims and objectives of the plan generally designed for 10 years.
- Record past history of the forests and the regions.
- Brief outline of climate, rainfall, geology, rock and soil, temperature and many other locality factors.
- Forests and forest types, botanical resources.
- Statistics of growth, yield volume of various crops various. damages suffered by the crops.
- Various management system practices.
- Enumeration of crop.
- Wildlife resources.

The foresters also prepare working plans and codes and working plan manual for standardising the schedule of preparation of a working plan. The state of Uttar Pradesh was pioneer in the matter of working plan organisation. Trevor and Smythie's hand book, "Practical Forest Management" details the method of Forest Management. Source of the classic working plans are :

Hein's working plan of Singhbhum Forests (1890).
Manson's working plan for Darjeeling Forests (1892).
Heining's working plan for Sundarbans (1893).
Caccia's working plan for Hosangabad (1894).
D' Arey's working plan for Kaleswar (1892).
Fisher's working plan for Kulu (1894).

The importance of proper working plan organisations with adequate staff was realised in the beginning of last century. The plan writers kept close liaison with silviculture research. During the period of World War-II (1939-'45) the working relating to working plans suffered seriously. Burma and the Andamans suffered the most.

Mr. Munro, the Conservator of Forests, Travancore, was the first officer to prepare forest working plan on the 20's and 30's of 19th century. From a study of growth factor of teak he forecast in 1837 that 1,00,000 trees in Travancore were fit for felling; later Cleghorn, Tremenheere, O' Brien and others worked in growth figures of teak. But accurate growth figure was collected by Dr. Brandis by his linear surveys. They also worked to safeguard overfelling of trees. He was assisted by Schleich and Ribbentrop and then proper working plan took its first stride.

Why they Valued Working Plans?

The colonial foresters laid the foundation of scientific management, handle of which was the working plan.

Alas! the authors were disappointed to locate such plans prepared for the country – whose total number may exceed 3000 in Indian libraries not even in the National Library. I think it is a poor recognition of the importance of such plans even by the foresters.

The readers may like to have a look at a small wooden cubical in the Commonwealth Forestry Institute, Oxford, U.K. and find hundreds (may be thousands) of copies of the working plans prepared for all the forest divisions of India. A reader may gather knowledge about Indian Forestry spending some time in that one cubical; such wonder is the collection.

Incidentally, the authors may cite the saying of the famous writer of International fame Late Nirad C. Chowdhury. One Indian asked him why has he made Oxford his home and not India. His reply was that his fields of study includes Sanskrit literature concerning India, the literatures of which were available in one Ballial Library at Oxford which he would not even find if he searched for such documents travelling from Kanyakumari to Kashmir and Gujarat to Assam.

Working plans are invaluable documents on Indian land management and socioeconomic structure of Indian people.

Forest Survey of India's Contribution

Scientific Forestry took a very positive shape and direction when the Government of India established a Department (1965) called "Preinvestment Survey of Forest Resources" with assistance from U.N.D.P. Use of aerial photograph, satellite imagery interpretation, computer data processing, reconnaissance survey by helicopter and the like was made to assess the resources of tree and Bamboo vegetation. This department was transformed into Forest Survey of India in 1981.

Advanced techniques of sampling, surveying, photogrammetry and data evaluation were applied in resource survey. U.N.D.P. assistance terminated in 1969 and the Indian experts took the charge of sampling.

These two departments were pioneer in Indian Forestry in the assessment of the following which must be considered a unique achievement although foresters' adopted various sampling methods in survey of resources since early years of forestry :

- Assessment of total growing stock of an area with an error of 10 percent at 95 percent probability level.
- Sampling techniques for various types of forests were determined.

- Inventory designs were made for assessing resources of various forest types.
- Local volume equation was made for each species.
- Biomass studies.
- Density of trees per ha.
- Distribution of diameter class of various species.
- Assessment of regeneration status of species.
- Assessment of rottenness of timber.
- Assessment of volume for matchwood, plywood, furniture wood, construction wood, etc.
- Assessment of forest cover density wise.
- Assessment of depletion of forests.

The F.S.I. can now confidently say how many trees of some particular tree occur per ha and what are their diameter class. This may outline an evolutionary trend of the species and indicate biotic interference in the forests.

Of late F.S.I. has extended their inventory in village areas outside the Reserve Forest.

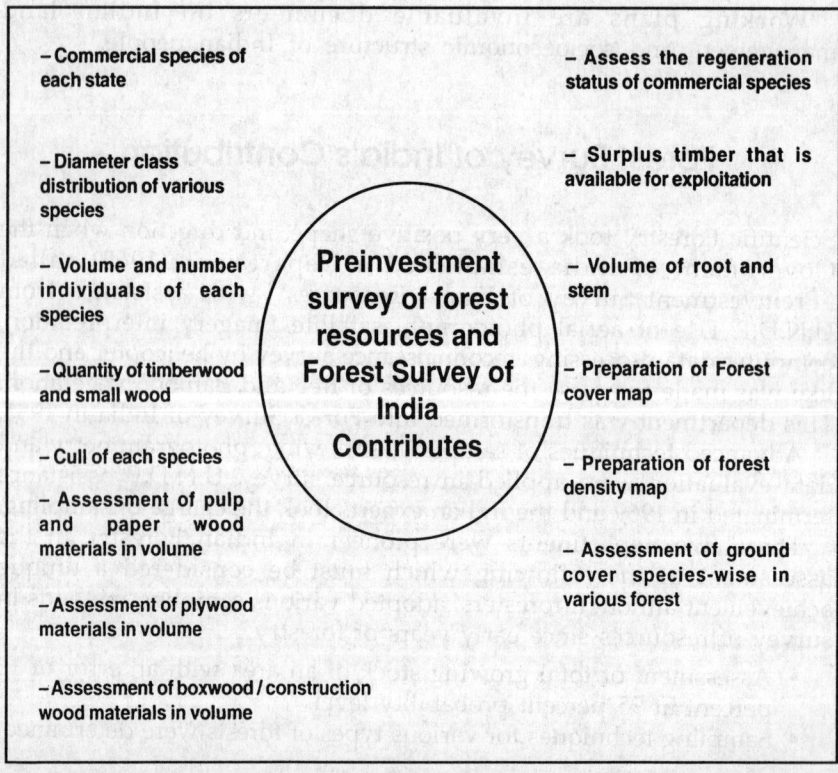

Industrial potential of various species with enumeration down to 10 cm diameter class were done all over the States of India. Valuable documents were compiled on the basis of which many industries grew in India.

Medicinal Plants
(A dwindling resource)

The authors sincerely feel that the vast reservoir of medicinal plant resources of India is in peril. The readers may peruse the chapters on 'medicinal plants' detailed in the books: (i) Biodiversity Endangered (India's Threatened Wildlife and Medicinal Plants) and (ii) Megadiversity Conservation (Flora, Fauna and Medicinal Plants of India's Hot Spots).

The authors have analysed various issues on the subject and have given a broad view of the status of important plants.

The authors have stressed the point about qualitative study by the Botanical Survey of India on the floras of various region. Without quantitative studies the status of any plant cannot be ascertained. As such, analysis of ground flora (herbs, shrubs, climbers) which form the bulk of medicinal plants is essential to determine the status of a plants.

Forest Survey of India, has initiated the study of ground flora all over the country which should focus some important features of ground flora.

Research

In the interim report of the National Commission on Agriculture the following suggestions were made indicating the fields of research:

* Forestry and biological research.
* Industrial and utilization research.
* Forest Management and operational research (including statistics, economics, marketing).

Since the beginning of twentieth century forestry research has taken up a regular research schedule in various departments which are –

Timber mechanics, wood seasoning, timber engineering, wood anatomy, forest botany, forest pathology, forest entomology, wood preservation, wood seasoning, celluloid and paper, chemistry of forest product and several other plants and fungi specimens have been kept in herbaria for research.

In 1987, Indian Council of Forestry Research and Education (ICFRE) was created under the Ministry of Environment and Forests. Later it was constituted into an autonomous body. ICFRE is headed by a Director General, who is assisted by a Director, Research and Director, Finance. It has under institutes and centres such as:

Forest Research Institute, Dehradun; Institute of Tropical Forests, Jabalpur; Institute of Arid Zone Forests, Jodhpur; Institute of Rain and Moist deciduous Forests Research, Jorhat; Institute of Genetics and Tree Breeding, Coimbatore; Institute of Wood Science Technology, Bangalore; Conifer Research Centre, Simla and Advance Centre for Forest Productivity, Ranchi.

Outstanding research finding have been achieved by the following : (only few have been mentioned):

- Dr. Gamble in Wood anatomy.
- Dr. Brandis, Kanjilal, Parkinson, Haines, Kurz, Raizada, Sahni and others in the field of Botany.
- Dr. K. A. Chowdhury, Dr. S. S. Ghosh in the field of Wood Technology.
- Dr. Stebbing, Dr. Chatterjee and Dr. Beeson in the field of Insect and Insect ecology.
- Dr. H. G. Champion, Dr. Trever and Dr. Troup in the field of Silviculture.

To mention few more names which are:

Silviculture	Botany
Dr. Troup	R. N. Parker
E. Marsden	C. E. Parkinson
S. H. Howard	B. L. Gupta
H. G. Champion	K. D. Bagchi
J. C. Sengupta	M. B. Raizada
C. G. Trevor	K. C. Sahani
M. V. Laurie	**Entomology**
P. N. Deogon	A. P. Imms
Utilization	C. F. C. Beeson
H. Troller	E. P. Shebbeny
C. C. Wilson	**Paper Pulp**
B. S. Chengappa	M. P. Bhargava
M. P. Bhargava	W. Raitt
Timber Testing	**Timber Seasoning**
Limaye	S. N. Kapur
Wood Technology	M. A. Rehman
R. S. Pearson	
Dr. K. A. Chowdhury	
Dr. Brown	
Dr. S. S. Ghosh	

SOME APPLIED FIELDS OF RESEARCH
(VARIOUS SPECIES EXPERIMENTED IN INDUSTRIES)

Pencil
Juniperus virginiana
J. procera
Cupressus torulosa
Cedrus deodara
Alstonia scholaris
Kydia calcycina
Lophopetalum whightianum

Battery Separator
Cupressus torulosa
Pinus wallichiana
Cedrus deodara
Adina cordifolia
Pulp and paper
Eucalyptus sp.
Bamboo sp.
Boswellia serrata
Acacia sp.
Katha
Acacia catechu
Ammunition box
Mangifera indica
Boswella serrata
Sports Goods
Morus alba
Celtis australis
Fraxinus sp.
Juglans regla
Toona ciliata
Melia azadirach
Morus alba
Mangifera indica

Some Forest Based Industries

Matchbox
Sideroxylon logepetiolatum
Endospermum malaccense
Canarium euphyllum
Bombax ceiba
Anthoceplalus chinensis
Bowellia serrata
Hymenodictiom excelsum
Sterculia villosa
Plywood
Terminalia bellerica
Schima wallichii
Dipterocarpus macrocarpus
Shorea assamica
Rosen and Terpentine
Pinus roxburghii
P. kasiya
Mathematical instruments
Cupressus torulosa
Pinus wallichiana
Juglans regia
Aesculus indica
Dalbergia latifolia
Tectona grandis
Adina cordifolia
Toona ciliata
Pterocarpus marsupium
Gardenia latifolia
Textile auxiliaries
Adina cordifolia
Mitragyna parviflora
Gardenia latifolia
Betula sp.
Michelia champaca
Gmelina aborea
Toona ciliata
Fagara budrunga
Amoora wallichii
Chukrasia velutina

Education and Training

FOREST EDUCATION AND TRAINING

The Colonial administrators had tremendous foresight and vision. They created forest services at par with civil and police services. As such keeping the special nature of duties of a forester in view, who has to perform a very broad based multifaceted education and training, a composite programme was created.

Some special features of I.C.F.R.E. (Indian Council for Forestry Research and Education) in regard to its some rich collections are enlisted below :

- Forest pathology branch has a collection of more than 10,000 specimens of plant diseases and about 900 fungal isolates.
- Forest entomology branch has more than 20,000 identified specimens of insects.
- Wood anatomy branch has over 10,000 specimens of various species.
- Botany herbarium has more than 2.5 lakh plant specimens.

Central Government Forest Courses.

- Post Graduate and post doctorate research at the Forest Research Institute.
- It has been constituted into a deemed University.

Indian Forest Service candidates are recruited through the Union Public Service Commission who select candidates through all India examination. They are treated as salaried probationers and trained at Indira Gandhi National Forest Academy, Dehradun. The probationers are required to undergo a foundation course of 4 months at the National Academy of Administration at Mussourie along with IAS and IPS probationers.

The State Forest Officers are selected by the States and sent to State Forest Colleges at Dehradun or Coimbatore or Barnihat (in Assam).

Forest Education

The Colonial foresters had the determination to run the forest administration in India on a firm scientific footing. So they recruited three German Officers – Sir D. Brandis, Dr. Schleich and Dr. Ribbentrop who were well trained in forestry to set up an organization for training of European and Indian Foresters. In formulating various categories of officers for training at different levels and in selection of syllabii one can find their wide visionary approach in this field.

The training course for the Forest Rangers and Foresters started in 1878 with Col. Bailey as Director of School in Dehradun. The training of Indian Forest Service probationers lasted up to 1932. In 1938 a decision was taken to reopen the college for training of the new superior provincial service officers. Meanwhile the Rangers College which was closed in 1933 was reopened in 1935. The Rangers College at Coimbatore which was started in 1912 by the Government of Madras was taken over by the Central Government.

Education started for 4 categories of service personnel which are:

- I.F.S. Superior Service Officer.
- Range Officers.
- Foresters/Beat Officers.
- Forest Guards.

I.F.S. Officer's training was fixed in Dehradun. Forest Ranger's Training College was bifurcated into two – N.F.R.C. (Northern Forest Rangers College) at Coimbatore. Almost all the states having their own institutes for training of foresters and forest guards.

During 1932–38 many states sent their officers to Edinburgh.

Spread of Forestry Education in Agricultural Universities

Several Agricultural Universities have included the following subjects of study in their curricula:

- Tree Forest Improvements.
- Forest soil, fertilization and irrigation.
- Forest tree propagation and nursery management.
- Weed Control.

Forestry being multidisciplinary venture, a very wide range of subjects were taught.

Some eminent features in forestry educators were : R.B. Cornwell (1925-29), E. C. Mobb (1930), D. Davis (1931), W. T. Hall (1937), C. R. Ranganathan (1938), S. A. A. Arvery (1943-47). S. S. Mandal, A. B. Rudra, N. J. Masani and others.

The syllabus covered a wide range of subjects as the foresters had to have a hostile approach needed for this complicated broad-base profession.

FORESTRY IS A MULTIDISCIPLINARY SUBJECT

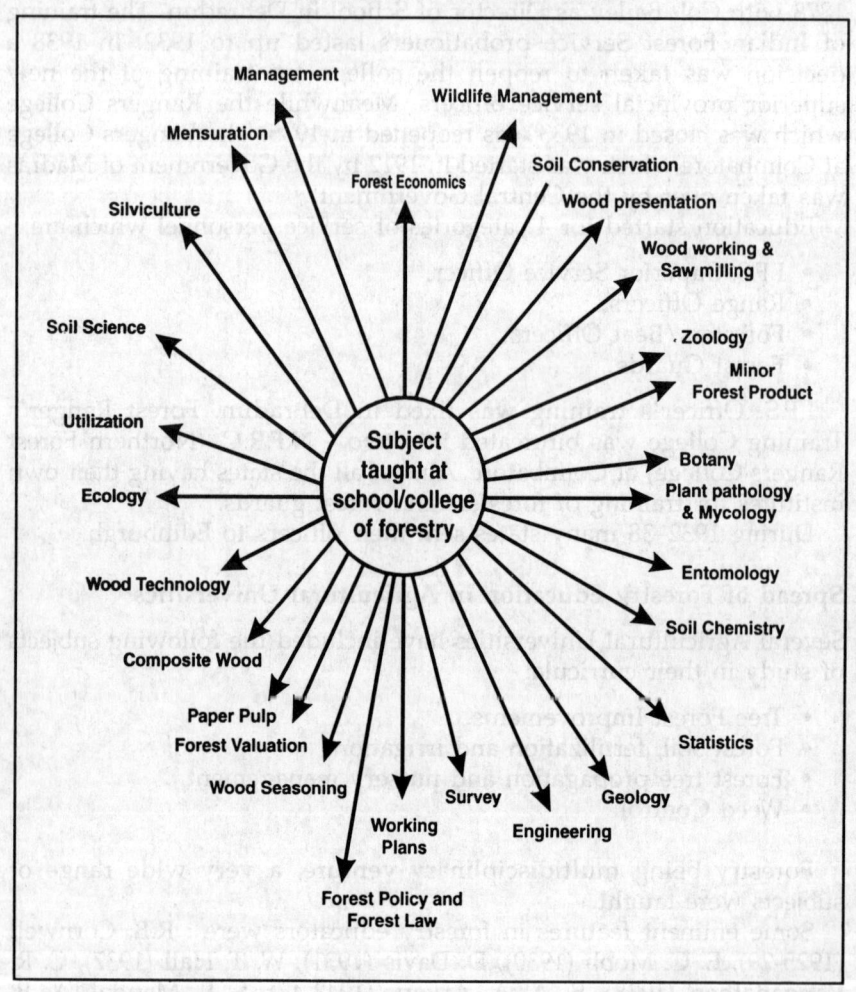

RIGORS OF TRAINING MADE THE FORESTERS A TOUGH AND DISCIPLINED CIVIL OFFICER

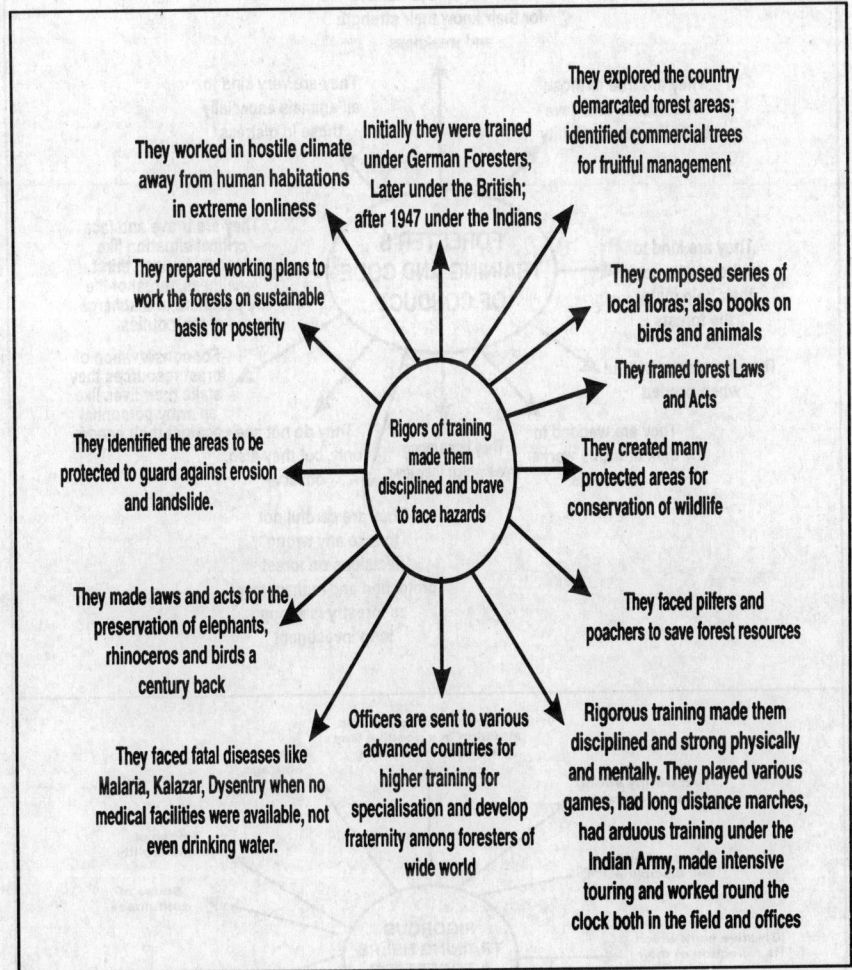

They explored the country demarcated forest areas; identified commercial trees for fruitful management

Initially they were trained under German Foresters, Later under the British; after 1947 under the Indians

They worked in hostile climate away from human habitations in extreme lonliness

They prepared working plans to work the forests on sustainable basis for posterity

They composed series of local floras; also books on birds and animals

They framed forest Laws and Acts

They identified the areas to be protected to guard against erosion and landslide.

Rigors of training made them disciplined and brave to face hazards

They created many protected areas for conservation of wildlife

They made laws and acts for the preservation of elephants, rhinoceros and birds a century back

They faced pilfers and poachers to save forest resources

They faced fatal diseases like Malaria, Kalazar, Dysentry when no medical facilities were available, not even drinking water.

Officers are sent to various advanced countries for higher training for specialisation and develop fraternity among foresters of wide world

Rigorous training made them disciplined and strong physically and mentally. They played various games, had long distance marches, had arduous training under the Indian Army, made intensive touring and worked round the clock both in the field and offices

CODE OF CONDUCT WAS A PART OF TRAINING

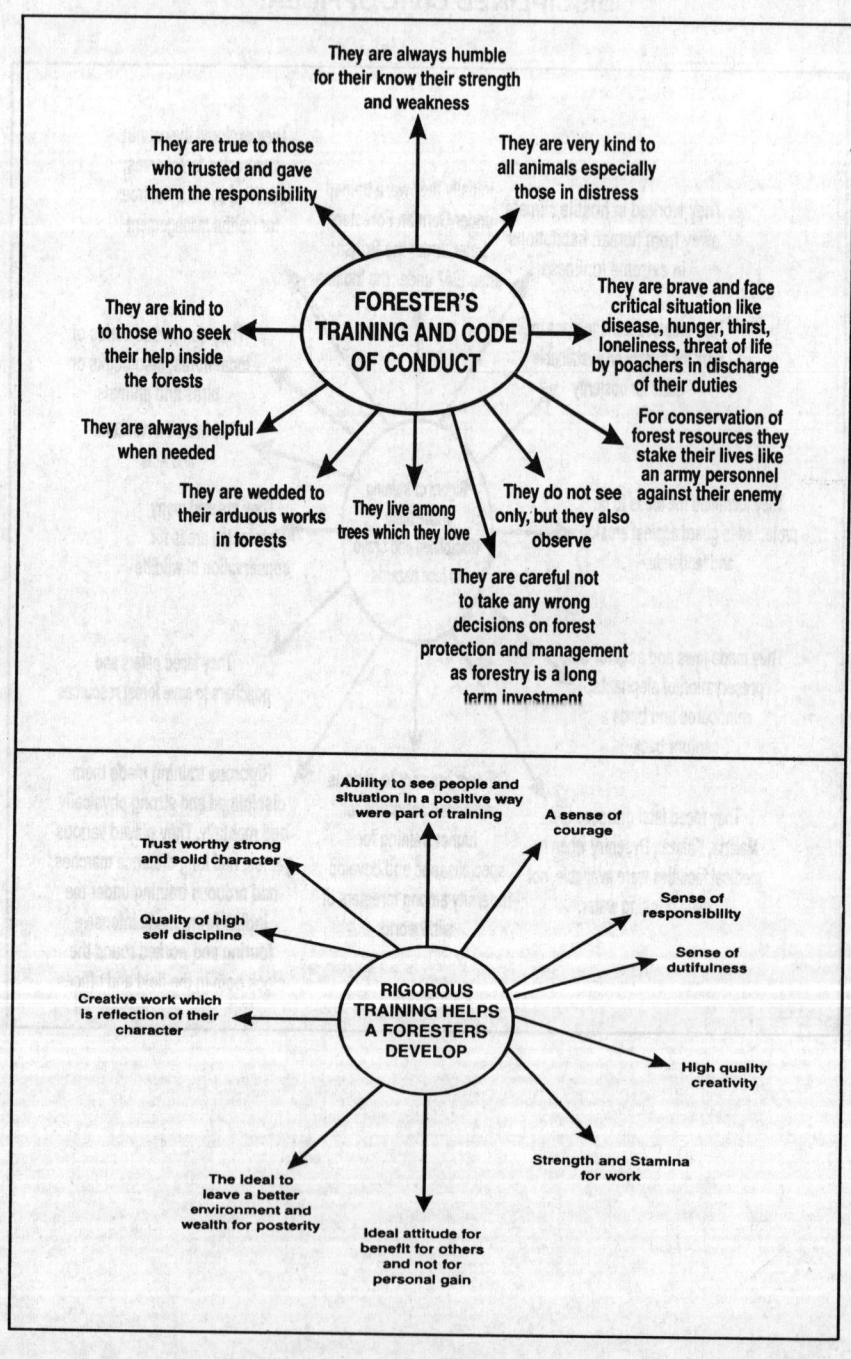

They are always humble for their know their strength and weakness

They are true to those who trusted and gave them the responsibility

They are very kind to all animals especially those in distress

FORESTER'S TRAINING AND CODE OF CONDUCT

They are kind to to those who seek their help inside the forests

They are brave and face critical situation like disease, hunger, thirst, loneliness, threat of life by poachers in discharge of their duties

They are always helpful when needed

For conservation of forest resources they stake their lives like an army personnel against their enemy

They are wedded to their arduous works in forests

They live among trees which they love

They do not see only, but they also observe

They are careful not to take any wrong decisions on forest protection and management as forestry is a long term investment

Ability to see people and situation in a positive way were part of training

A sense of courage

Trust worthy strong and solid character

Sense of responsibility

Quality of high self discipline

Sense of dutifulness

RIGOROUS TRAINING HELPS A FORESTERS DEVELOP

Creative work which is reflection of their character

High quality creativity

Strength and Stamina for work

The Ideal to leave a better environment and wealth for posterity

Ideal attitude for benefit for others and not for personal gain

HONEY COMB TRAINING INSTITUTE

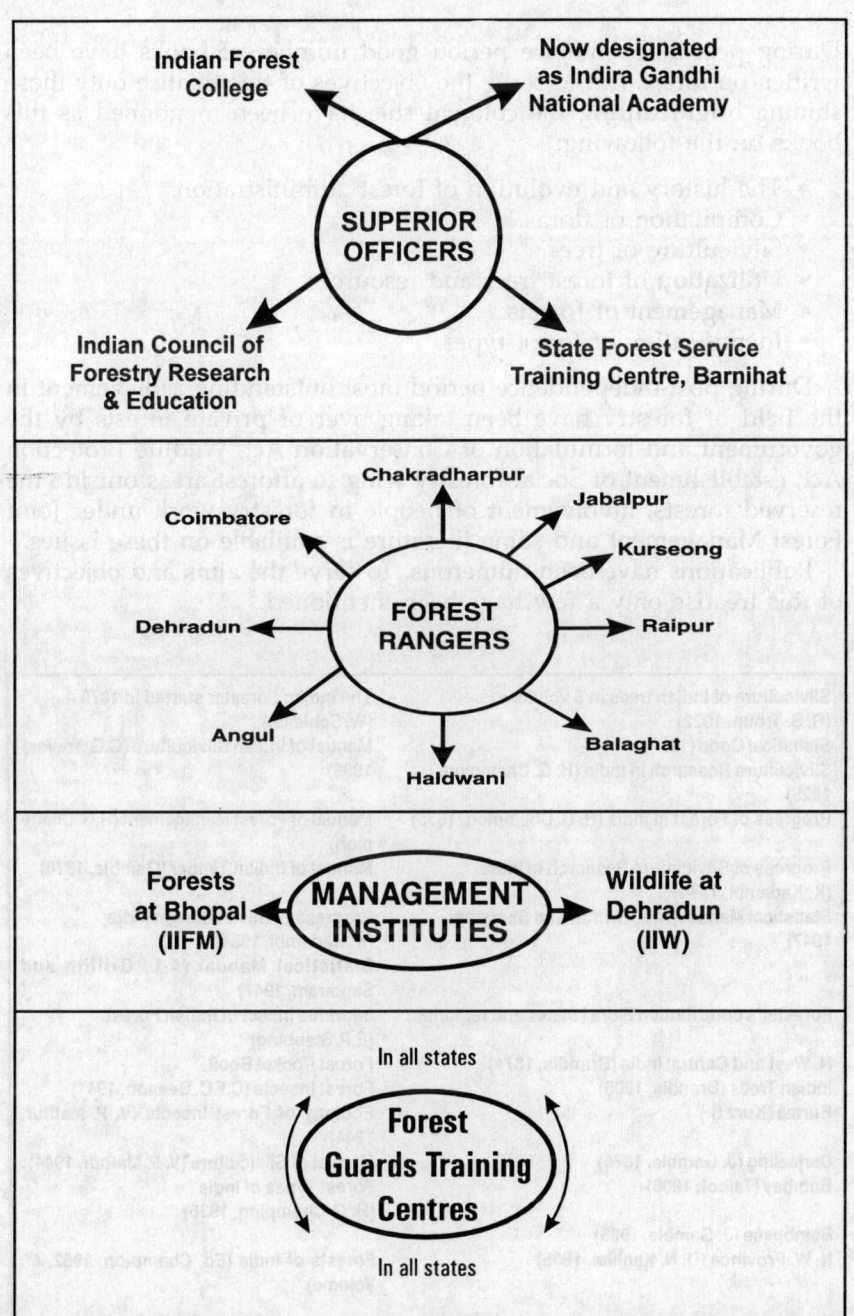

Outstanding Publications

During post-independence period good number of books have been written on forest. But to serve the objectives of this treatise only those shining bright during the colonial rule have been mentioned as this books on the following:

- The history and evolution of forest administration.
- Compilation of floras.
- Silviculture of trees.
- Utilization of forest trees and resources.
- Management of forests.
- Identification of forest types.

During post-independence period most outstanding achievement in the field of forestry have been taking over of private forests by the government and formulation of Conservation Act, Wildlife Protection Act, establishment of Social forestry wing to afforest areas outside the reserved forests, involvement of people in forestry work under Joint Forest Management and some literature is available on these issues.

Publications have been numerous. To serve the aims and objectives of this treatise only a few have been mentioned.

Silviculture of Indian trees in 3 volumes (R. S. Troup, 1922)	The Indian Forester started in 1875 (W. Schleich)
Statistical Code (1921)	Manual of Indian Silviculture (C.G. Trevor; 1938)
Silviculture Research in India (H. G. Champion, 1929)	
Progress of Forest in India (H. G. Champion, 1935)	Manual of Forest Management (H.G. Champion)
Progress of Silviculture Research in India (K. Kadambi, 1954)	Manual of Indian Timber (Gamble, 1878)
Statistical Manual (A. L. Griffith and Santaram, 1947)	Progress of Silviculture in India (K. Kadambi, 1954)
	Statistical Manual (A.L. Griffith and Santaram, 1947)
Forester's contribution Flora (States and regions)	Injurious insect of Indian Forest (E.P. Stebbing)
N. West and Central India (Brandis, 1874)	Forest Pocket Book
Indian Trees (Brandis, 1906)	Forest Insects (C.F.C. Beeson, 1941)
Burma (Kurz S.)	Ecology of Forest Insects (V. P. Mathur, 1944)
Darjeeling (J. Gamble, 1875)	Manual of Silviculture (V. P. Mathur, 1944)
Bombay (Talbot, 1906)	Forest Types of India (H. G. Champion, 1936)
Bambusae (J. Gamble, 1925)	
N. W. Province (U. N. Kanjilal, 1935)	Forests of India (Ed. Champion, 1962, 4th Volume)

Andamans (Parkinson, 1935)	Commercial Timber of India (R. S. Pearson, 1932)
Bombay (Cook 1931)	Manual of Forest Utilization (H.Trotterz, 1940)
Madras (J. Gamble, 1931)	Commercial Timber of India (K. A. Chowdhury, 1927)
Assam (Kanjilal *et al* 1934-41)	All India Yield Table for Teak (M. V. Laurie, 1937)
	Manual of Indian Trees (N. L. Bor, 1953)
	The Forest of India [E.P. Stebbing (3 vols)]

Prominent post-independence publications :

Forest and Forests – K. P. Sagreiya (1967).
Indian Forests and Forest Products Terminology (1965).
Revised Forest types of India – H. G. Champion and S.K. Seth (1968).
Lessons in Forest – M. L. Sharma (1959).

Publication of History of Indian Forests

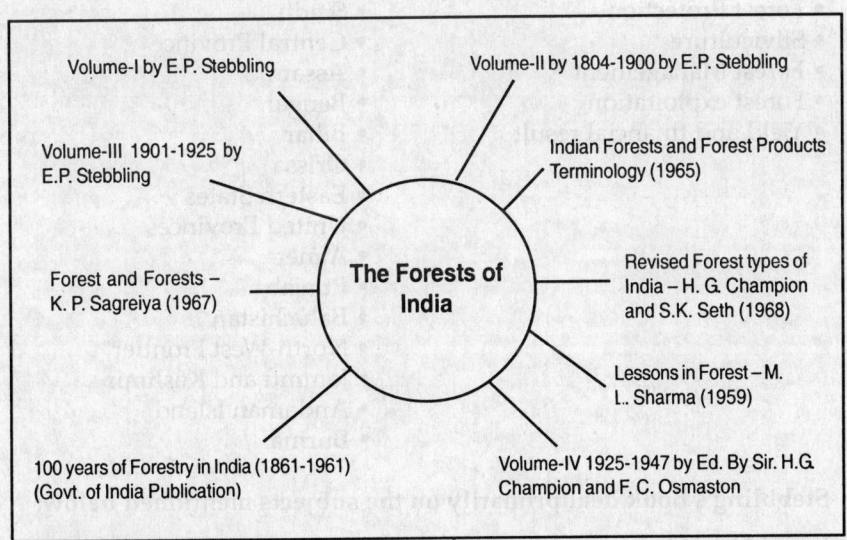

Volume-I by E.P. Stebbling

Volume-II by 1804-1900 by E.P. Stebbling

Volume-III 1901-1925 by E.P. Stebbling

Indian Forests and Forest Products Terminology (1965)

Forest and Forests – K. P. Sagreiya (1967)

The Forests of India

Revised Forest types of India – H. G. Champion and S.K. Seth (1968)

Lessons in Forest – M. L.. Sharma (1959)

100 years of Forestry in India (1861-1961) (Govt. of India Publication)

Volume-IV 1925-1947 by Ed. By Sir. H.G. Champion and F. C. Osmaston

Book by E. P. Stebbling (1900-1925) deals with the following subjects:

- Review of administration.
- Progress of administration.
- Training of probationers.
- Inauguration of forest research institutes.

- Forest protection – Fire.
- Progress of silviculture.
- Irrigated and non-irrigated plantation in Punjab and UP.
- Progress of working plans.
- Record of forest dept. during World War 1914-19.
- Progress of modern method of exploitation.
- Progress of Yield and revenue.
- Aerial forest survey in Burma.

Champion and Osmaston's deals with subjects and spread of administrative control of forest as detailed below:

General Progress in Forestry

- Review of General Administration and all-round progress of administrative and activities
- Forest administration, Policy, Law
- Forest Research Institute and progress in Forest Research
- Forest Protection
- Silviculture
- Forest Management
- Forest exploitation
- Yield and financial result

- Forest Administration spread all over India

- Madras Presidency
- Bombay Presidency

- Sindh
- Central Province
- Assam
- Bengal
- Bihar
- Orissa
- Eastern States
- United Provinces
- Ajmer
- Punjab
- Baluchistan
- North-West Frontier
- Jammu and Kashmir
- Andaman Island
- Burma

Stebbling's book deal primarily on the subjects mentioned below:

Forest Organisation and l Methodica System of Managemen	– Progress of Forest Conservancy and the Development of Forest Department (1871-1900)
Training of Forest Probationers (Institute of Continental training by Brandis)	– Progress of forest administration – Education of staff – In forest protection – Formation of Plantation

– In forest protection
– Improvement of forest crop
– Growth of Forest working plans

Progress of Forest Administration
– Madras Presidency, Coorg and Mysore
– Bombay Presidency
– Burma
– Central Province
– Punjab
– Northwest Province and Sindh
– Himalayan Province
– Bengal and Assam

11

Forester's Unique Contribution in Conserving Biodiversity and Megadiversity
— Global Forest Jewel in Northeast

For the scientific management of forests the erstwhile forester's formulated various principles for determination of the growing stock, various species composition and site qualities for enabling preparation of plans for individual areas according to the relevant parameters. Foresters prepared working plans separately for every division, based on the principles of sustained yield. In this process they also became the able custodians of the forest resources along with wildlife. For the purpose of conservation and preservation of forests and wildlife, various national parks and wildlife sanctuaries have been created along with the pro-people approach of joint forest management. The authors feel that the wildlife and medicinal plant resources of the country are very much under threat.

Forester's Unique Contribution in Conserving Biodiversity and Megadiversity

— Global Forest Jewel in Northeast

The present authors endorse the note of warning about uncontrolled exploitation of forests all over the country.

While presenting a paper cutting (*The Telegraph* dt. 21.02.2005) the authors noted that the richness of flora and fauna of N.E. India from Sikkim-Darjeeling Himalayas eastwards has been expressed in various literatures and also recognized by various world organizations.

The bibliography of this book mentions a number of books initially under the authorship of A.B. Chaudhuri, the author and later under the authorship of A. B. Chaudhuri and D. D. Sarkar, co-author categorically mentions about the important biogeographical situation of the region. In recent years their books on "Wildlife and ground flora", "Biodiversity endangered" and particularly "Megadiversity conservation" present facts and figures about the richness of flora and fauna of the region and other relevant problems of the area.

Some extracts of this book may be very relevant as the authors take the pride to focus the issue in 2002, which some world organizations record about the creditable jobs of foresters in creating protected areas for their conservation.

Global Forest Jewel in Northeast

The Northeast has the second richest forest reserve in the world in terms of plant diversity, which the World Wide Fund for Nature (WWF) has revealed after an extensive study.

The area surveyed by the WWF is called the North Bank Landscape, spanning 3,000 sq. km of the Himalayan foothills north of the

Brahmaputra river in Assam and parts of Arunachal Pradesh, north Bengal and Bhutan.

"Preliminary results indicate that the North Bank Landscape may be surpassed only by the Sumatran forests in Indonesia, thus making it the second riches centre of plant diversity in the world," the WWF said in its report, which was released during the centenary celebrations of Kaziranga National Park in February '05.

All 3,000 sq.km of forests were subjected to a "rapid appraisal" under the Asian Rhino and Elephant Action Strategy programme of the WWF. The surveyors found and extraordinary number of plant species, 107, in a single 200-square metre plot.

The WWF is convinced at richness of the forests in the area is higher than similar lowland forests in other biodiversity hotspots like Brazil, Cameroon, New Guinea and Peru.

The study was coordinated by WWF-India and WWF Asian Rhino and Elephant Action Strategy (AREAS) project in association with the Centre for Biodiversity Management, the Smithsonian Institution, the National Zoological Park and the Conservation Research Centre with additional funding by the MacAurthur Foundation.

In December 2003, two teams from four Indian institutions undertook an initial survey within the eastern Himalayas, including the proposed KaSoPaNa reserve, comprising the Kameng-Sonitpur, Pakke and Nameri forests on the Assam-Arunachal border.

Adrew Gilson, the author of the report and head of the Centre for Biodiversity Management, described the North bank Landscape as "extraordinary" and "extraordinary" and a "jewel in the crown of Indian forests".

"Uncontrolled exploitation of forests and eviction of large mammals is clearly impacting both plant and animal habitats – a situation that is unlikely to improve with increasing elephant-human competition for common resources. Appropriate policy intervention at national and international levels is urgently needed to establish a regional framework for ensuring human livelihood and balanced conservation management", the WWF report said.

[*The Telegraph* dt. 25-02-2005]

The following books, authored by A. B. Chaudhuri and D. D. Sarkar deal with some subjects, very relevant to the present issues.

Biodiversity Endangered (India's threatened wildlife and medicinal plants – 2002).

Definitions, concepts values, etc. of biodiversity, the crisis faced by the medicinal plants of India, a quantitative assessment of medicinal plants, threatened wildlife of India, protected area and sustainable development.

Megadiversity Conservation (Flora, Fauna and Medicinal plants of India's Hot Spots – 2003).

The book 'Biodiversity endangered' discusses specifically the richness of flora and fauna of North-East India under the parameters.

1. Biodiversity and Conservation

  1.1.   Introduction
  1.2.   Definitions
  1.3.   Global Biodiversity Assessment
  1.4.   Biodiversity Concept
  1.5.   Valuing Biodiversity
  1.6.   Sustainable development : Need for valuing resources
  1.7.   Biodiversity and Ecosystem processes
  1.8.   Measuring Biodiversity
  1.9.   Conservation : Approaches and Researches
  1.10. Biodiversity in various living organisms
  1.11. Restoring Biodiversity
  1.12. A World Bank Guidelines for monitoring and evaluation for biodiversity projects

2. Quo Vadis, Medicinal Plants

  2.1.   Floral and Faunal resources — Face a crisis
  2.2.   Factors causing depletion of medicinal plants
  2.3.   Loss of Habitat
  2.4.   Background history of Ayurveda
  2.5.   Status report : Medicinal herbs
  2.6.   Status report : Medicinal shrubs
  2.7.   Status report : Climbing medicinal plants
  2.8.   Status report : Medicinal Tree Flora
  2.8.1. Some basic information
  2.8.2. Tree conservation
  2.8.3. Quantitative assessment of medicinal plant species
  2.9.   Medicinal plants : Lacunae in conservation and preservation
  2.10. Medicinal properties of some selected plant species
  2.11. List of Potential Drug plants
  2.11.1. -do- by R.N. Chopra
  2.11.2. -do- by Shibakali Bhattacharya
  2.11.3. -do- by Kirtikar and Bose

3. The Threatened Wildlife of India

  3.1.   India's Wildlife situation is under severe threat
  3.2.   Human population explosion vis-a-vis Wildlife
  3.3.   Conserving biodiversity resources
  3.4.   Checklist (tentative) and status of Mammalian fauna
  3.5.   Primates
  3.6.   Pangolins

Plantation Forestry

— Need for Afforestation of Denuded Lands and to Augment the Value of Forests

India has lost substantial areas of virgin forests in last 500 years and the pressure of a billion people is enough to deplete the forest cover all over the country. As such plantation forestry is the only remedy to maintain the ecological balance.

This book has been designed to serve and convince the intelligentsia that the foresters create and protect the human destiny and their prosperity and when a forest fails to regenerate naturally they come to the rescue. They artificially procreate plants by raising nursery, taking meticulous care and plant them to create the forest. So the foresters serves the posterity by providing them not only their requirement of fuelwood, timber, fodder, fruit, etc., they thereby create a salubrious climate and a pollution free landscape.

The forests during the process of development immediately attract ground and arboreal animals and birds, reptiles, lizards and other myriads of species of animals gradually taking shelter and food for living.

So by planting trees the foresters :

- *Gift wealth to the posterity,*
- *provide shelter to innumerable species of macro and micro animals,*
- *create a pollution free healthy climate,*
- *provide recreational sites for the tourists,*
- *create a natural laboratory and museum for students of science to work and learn,*
- *helps sequestering carbon dioxide,*
- *and many other direct and indirect benefits accrued from the forests.*

Plantation Forestry

— Need for Afforestation of Denuded Lands and to Augment the Value of Forests

Plantation Defined

A plantation is defined as "A Forest crop or stand raised artificially, either by sowing or planting." Artificial is sometimes considered 'man-made', though some confusion arises when a natural regeneration is enriched by planting. The following definitions may remove some confusion:

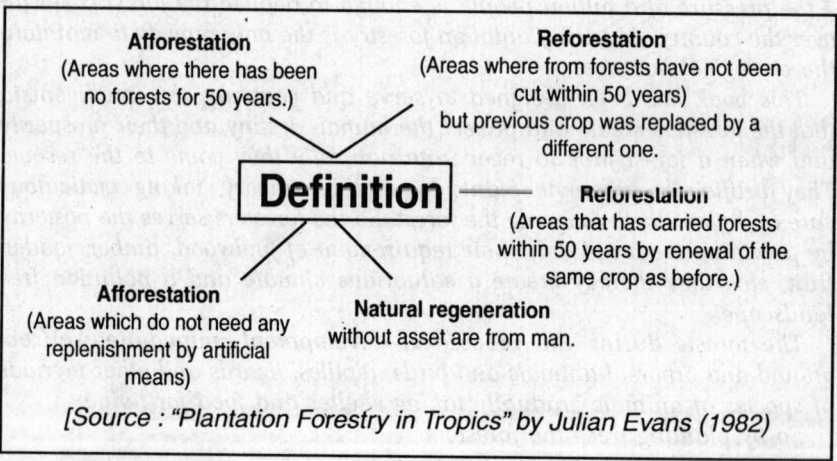

Afforestation
(Areas where there has been no forests for 50 years.)

Reforestation
(Areas where from forests have not been cut within 50 years) but previous crop was replaced by a different one.

Definition

Reforestation
(Areas that has carried forests within 50 years by renewal of the same crop as before.)

Afforestation
(Areas which do not need any replenishment by artificial means)

Natural regeneration
without asset are from man.

[Source : "Plantation Forestry in Tropics" by Julian Evans (1982)]

Plantations are distinct from natural forests because of their orderliness and uniformity, their linear rows of trees same age class and simpler management schedules. They have regular well-defined shape and boundaries.

Brazil, in spite of having a vast rain forest, embarked on a planting programme of half a million hectare per year, some countries exploited 90 percent of their forests for fuelwood and charcoal for cooking and heating. In the whole world, half of all wood harvested, is burnt for energy

India having 170 persons per sq. km. has millions of hectares (some say, about 13 per cent of its land area or 43.6 million ha. is waste and much of it is suitable for plantation forestry.

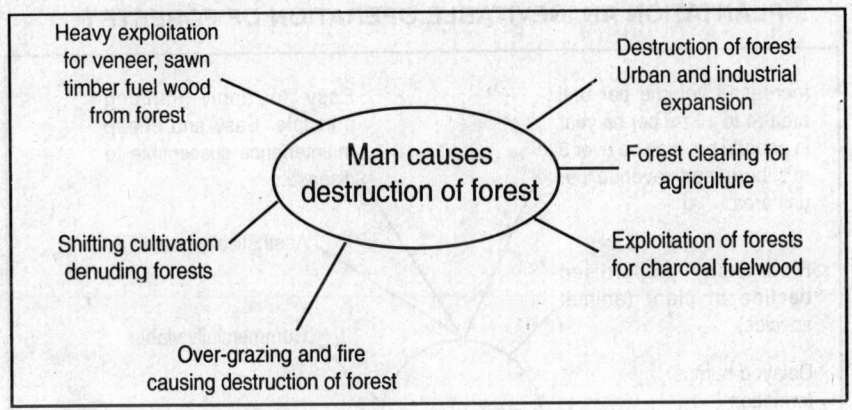

Principal Plant Species in Plantations –

Principal Species

Teak	**Sal**	**Eucalyptus**	**Acacia**
(Tectona grandis)	(Shorea robusta)	(Eucalyptus sp.)	(A. auriculiformis)

Bamboos
Dendrocalamus strictus
Bambusa vulgaris
B. nutans
B. arundinacea

PLANTATION

Other species

Casuarina equisitifolia (Jhau)
Gmelina arborea (Gamari)
Lagerstroemia speciosa (Jarul)
Acacia arabica (Babul)
Azadirachta indica (Neem)
Prosopis specigera
Acacia catechu
Cassia siamea
Albizzia sp.
Acacia leucocelola
Terminalica tomentosa
T. bellerica
Pterocarpus santalinus
Eucalyptus citriodora
E. Camaldulensis
E. teriticornis

Populus sp (Poplas)
Bombax ceiba
Dalbergia sissoo (shishum)
Anacardium occidentale
Pterocarpus marsupium (Bijasal)
Santalum album
Szygium cumini
Boswellia serrata
Xylia xylocarpa
Dalbergia latifollia
Anogeissus latifolia
Pongamia pinnata
Ailanthus excelsa

Pinus roxburghii
Cedrus deodara
Pinus smythiana
Cupressus torulosa
Cryptomeria japonica

PLANTATION AN INEVITABLE OPERATION OF FORESTRY

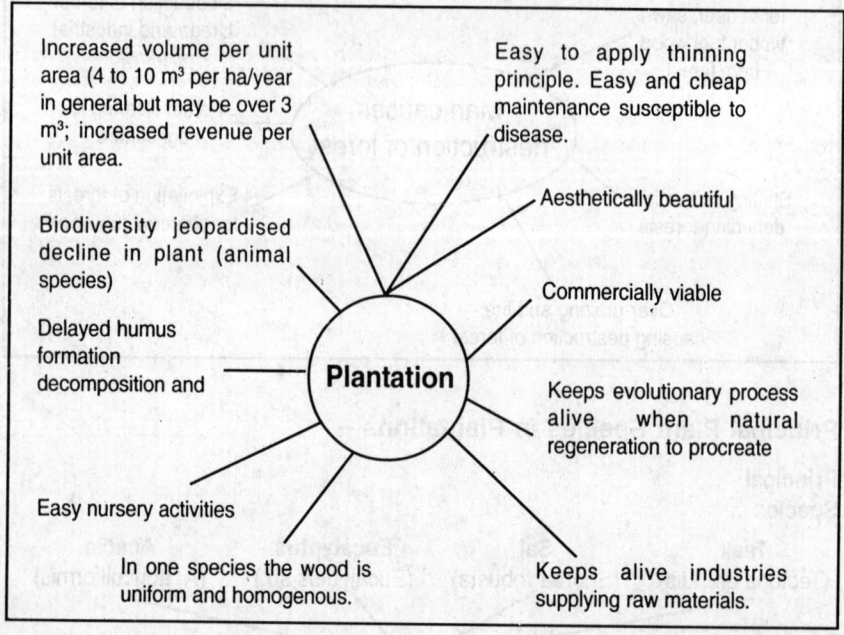

Increased volume per unit area (4 to 10 m³ per ha/year in general but may be over 3 m³; increased revenue per unit area.

Easy to apply thinning principle. Easy and cheap maintenance susceptible to disease.

Biodiversity jeopardised decline in plant (animal species)

Delayed humus formation decomposition and

Aesthetically beautiful

Commercially viable

Plantation

Keeps evolutionary process alive when natural regeneration to procreate

Easy nursery activities

In one species the wood is uniform and homogenous.

Keeps alive industries supplying raw materials.

Indian foresters take pride in creating plantation of innumerable species in all agro-climate zone. Bamboo, Teak, Sal, Sissoo, Casuarina, Eucalyptus, Acacia, Pine, Gamhar, are principal species of plantations.

IMPACT OF MANAGEMENT PRACTICES ON FOREST ECOSYSTEMS

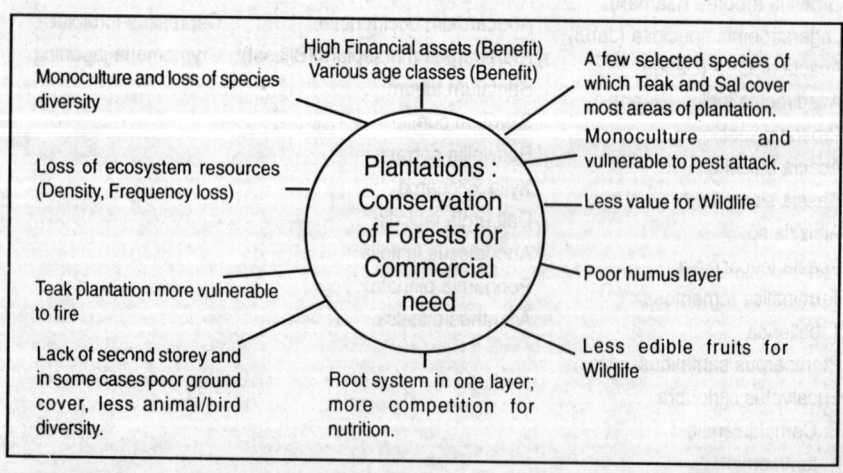

Monoculture and loss of species diversity

High Financial assets (Benefit)
Various age classes (Benefit)

A few selected species of which Teak and Sal cover most areas of plantation.

Loss of ecosystem resources (Density, Frequency loss)

Monoculture and vulnerable to pest attack.

Plantations : Conservation of Forests for Commercial need

Less value for Wildlife

Teak plantation more vulnerable to fire

Poor humus layer

Lack of second storey and in some cases poor ground cover, less animal/bird diversity.

Root system in one layer; more competition for nutrition.

Less edible fruits for Wildlife

Impact of Plantation

The history of forest management clearly shows that forests are mostly exploited commercially keeping all the conservation aspects well within view. Hilly areas were not clearfelled and are kept under protection Working Circle; areas where communication do not develop properly and there is no market kept under undeveloped working circle. When natural regeneration cease to establish, taungya method of artificially raising plantations introduced.

In Buxa Tiger Reserve, West Bengal, plantations were pure and mixed and by 1980 more than 25,246 ha. of forest areas were brought under plantation out of total area of 75900 ha. Similar situation exists in all the states.

The following facts may be perused prior to analysing the impact of conversion of high forests into plantations if :

- Bulk area was under grassy savannah-like formations.
- Plenty of areas had poor stock of trees.
- Substantial areas were under kukat trees, i.e. useless miscellaneous tree crop.
- Valuable trees other than sal was occurring sporadically and their abundance was limited that necessitated increasing the density of those species.

This situation and demand for some specific timber initiate raising of high value plantations all over the country. Since 1960 onwards demand for quick growing match wood, plywood, box wood and paper pulp wood species grew and various species were introduced in plantation schedule.

Areas brought under plantations after clear felling and burning and taungya cultivation had brought some direct impact on biodiversty as follows :

- Plantation areas were cleared of all weeds, i.e. herbs, shrubs, climbers and tree regenerations, on sustained basis.
- Plantation clearing, climber cutting and thinning had a direct impact on the ground vegetation which were thinned and weeded out.
- Fodder species and fruit yielding species were cleared of the sites which affected the herbivores, availability of fodder and cover and fruits for frugivorous birds.
- No strips of forests were retained as buffer in between the plantations as a measure to reduce incidence of insect epidemic and also for camouflage and movement of wildlife.

Quantitative analysis and other observations made in this study adequately reveal the changes of ground flora due to plantations.

SOME RAMIFICATIONS OF HUMAN-INDUCED VEGETATION CHANGE

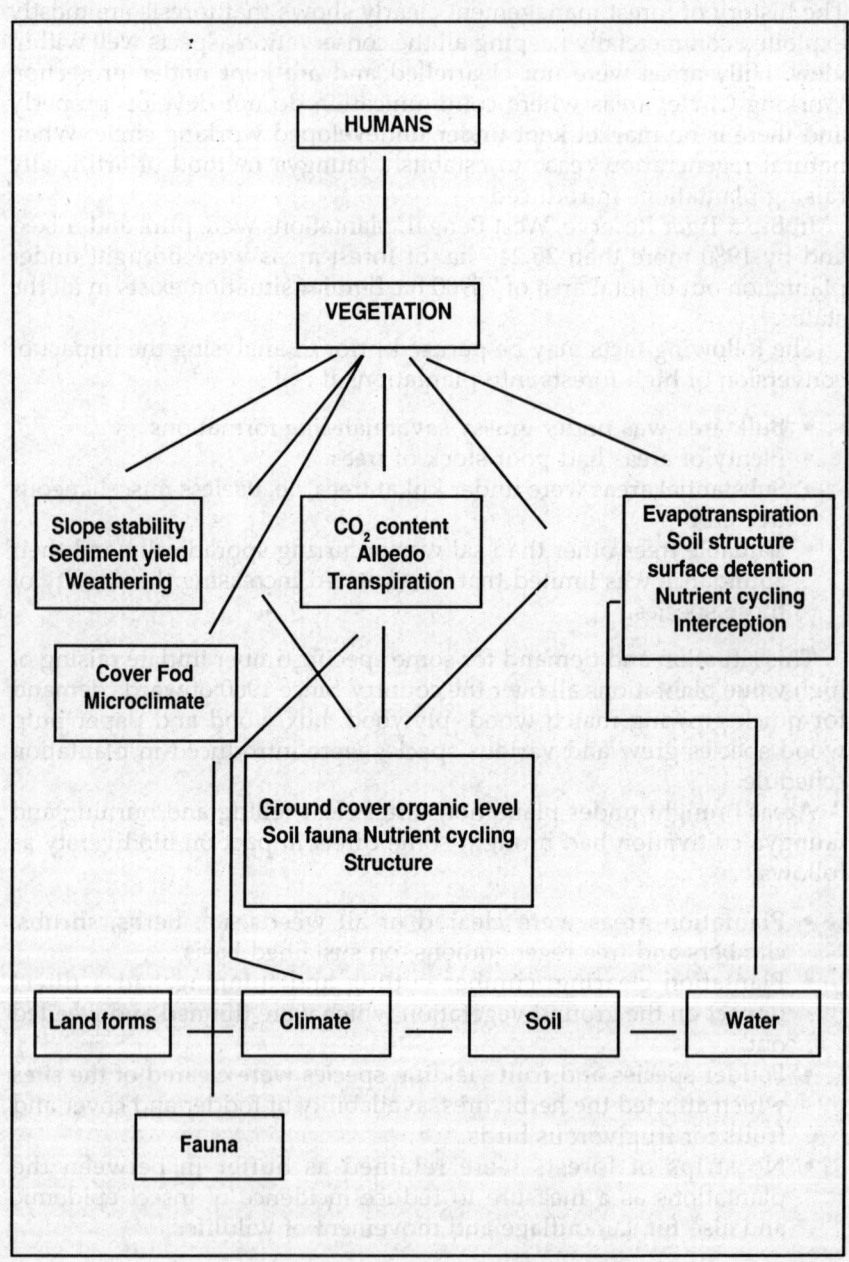

[*Source :* **The Human Impact on the Natural Environment by A. Goudia 1981**]

Beneficial Aspects of Plantations

- The value of crop increases substantially as production of timber per unit of area goes much high.
- The plantation crop represent various age classes like a normal forest.
- Management processes are easy as the crop is concentrated in blocks of a single species.
- Exploitation is easy.

Problems of Plantations

- The plantation ground flora has fewer species than high forest.
- Teak plantations have much fewer species than sal or other mixed plantations.
- Teak plantations have a very much thin humus layer; Sal and other sp. have a little better humus layer but much less than high forests.
- Teak plantations are vulnerable to fire hazard.
- Plantations have canopy layer of only a single crop.
- Root penetration in a monoculture plantation is generally at the same level which expose the roots to competition for nutrients.
- Monoculture plantations are more vulnerable to insect and fungal attack.
- The ecological concept of primary productivity - This concept illustrates an important differences between plantations and natural forests. In natural forests the rate of production of organic matter (gross primary productivity) is high and relatively little of it is stored (as wood); most is consumed in respiration and by heterotrophic organism such as animals, insects, fungi, here the forest are in a State of equilibrium. In plantation the rate of storage of organic matter (community production(is very high. But plantations do not reach equilibrium state where decay and breakdown balance new growth.
- There is very clear indication of poorer occurrence of ornitho fauna, mammalian and other fauna in plantations. There poor humus open ground cover, ecological niche and composite substrata make impossible for various animals to survive.

Evil effect of pure plantations of *Gmelina arborea, Swietenia mahagony, S. macrophylla, Bombax ceiba, Michelia, champaca,* etc. are well-known to West Bengal foresters. Epidemic in Teak plantations due to Hyblea purea and *Hapalia machaeralis* were menace in the past. (Net community productivity) is high. Clearfelling is done when maximum mean annual increment of stored organic matter (wood) culminates to maximize yield from the site.

The plantations do not grow to reach an equilibrium state where decay and break down balance new growth, but is cut to maximise wood yield from the site at regular intervals. In natural forest the system is practically closed–all that is produced is consumed within the ecosystem. So the plantation have a role in ameliorating the green house effect by fixing carbon, whereas the role of untouched forest is neutral (forest when cleared and burnt much carbon dioxide is released).

Organic Matter and Nutrient Cycling

Plantations are not ecologically dead but relatively poor (Evans, 1992)

- Nutrient intake in natural forests is relating constant from year to year with nutrients being rapidly recycled in the ecosystem. In plantations nutrients uptake varies with age of the stand.
- Plantations are less efficient at trapping released nutrients from decomposing litter than natural forests which has a maps of fine roots near the surface and further occupation of the whole rooting zone owing to the many different species present.
- Little organic matter accumulates beneath mature rain forest; much may build up under plantation.
- Plantation management may lead to significant nutrient loss from the site are harvested.

Habitat Diversity

There are variety of habitats in natural forests where consumers survive. One site one species will support fewer consumers. Dense canopy of one or few species do not allow a dense ground cover to grow. But there is a spatial diversity as plantations are of various age classes and cover a wide sites.

Plantations Considered as a Tool of Development (FAO, 1978)

Plantation forestry can significantly aid economic development.

Both the forestry sector policies of the World Bank (1978), Sector Policy Paper) and Asian Development Bank (1978, Proceedings of 8th World Forestry Congress by Richardson) endorse the important development potential of plantation.

The main benefits of plantation of forestry to the economic of developing countries are:

- Resource creation, to meet demand for wood and wood product.
- Development of a *Flexible resource to feed* fuelwood, timber, pulpwood, etc.
- Use of degraded land, otherwise useless for agriculture.
- Creation of employment in rural areas.
- Plantation establishment is a high level investment.

- Extensive plantation bring development of an infrastructures of roads, communication, services, houses, shops, schools, etc. often in remote areas.
- Environmental roles.

By planting trees, ground cover is re-established which will halt and eventually reverse the degradation of much of the tropical environment.

Plantations stabilize soil, prevent erosion, control run off in catchment areas, provide shelter from wind and heat, sand and dust.

Agro forestry is a key to sustainable land management.

Effect of Thinning

After thinning more light react beneath the canopy and there is usually a resurgence of weed growth, increase breakdown of litter, and sometimes epicormic shoots on the stem.

Also the *Watertable* may rise and the ground water since there is temporarily both less demand for soil moisture and less interception of rain by the canopy.

Lessening composition between trees has three main effects.

1. Lower natural mortality.
2. Remaining trees have deeper crown.
3. Increased crown expansion.

The effects on 2 and 3 above result in a greater photosynthetic areas on each remaining trees, thus increasing their growth.

Plantation and People

The purpose of growing trees are ecological, social and economic. Firewood, sawn timber, pulpwood, etc. are the main product available to the people, the plantations serve various emotional requirements and both the processes benefit the people. Plantation development brings benefits – better health, nutrition, education, greater access to goods and services and is, therefore, what people want. The impact on and response by the local community, not only in terms of enlarged employment prospect, cash income and new ambition, but also in cultural and social life cannot be ignored.

Impact

If the impact and disruptive effect of a development, such as "plantation project" is to be reduced, consultation and involvement of people must be sought. Unfortunately, where policies are based on economic growth, time and money are rarely allowed for this. Taungya method, reduce the impact of a plantation project.

The use of Tools in felling trees and raising plantation has changed over the years. Mechanisation of forest operation may sometimes be good economics, but does nothing to moderate the impact of development.

Impacts can be minimized in five possible ways:

- Consulting with and involving the local people.
- Employing those who are living in the fringes.
- Adopting practices to suit to local skills and abilities.
- Phasing development through many small-scale project.
- Using intermediate technologies not requiring highly specialised skills.

Total plantation are is 25,246 ha., of this, about 8578 ha. is covered by Teak and Jarul.

The authors bring to light that over the years of management following resources have suffered from various quantum of depletion:

- Loss of biodiversity.
- Depletion from grazing, which continued uncontrolled over years.
- Conversion of mixed crop into pure crop of few selected species.
- Lack of natural regeneration of various local trees, shrubs and herbs, etc.
- Over-exploitation of Canes, medicinal plants and some other species.
- Deterioration of physical and chemical properties of soil.
- Depletion of ground flora, fauna and soil due to introduction of Teak over large areas.

Habitat Improvement (including Soil moisture conservation and cane improvement)

1. This is an elaborate subject where many issues are involved, some of these issues are being mentioned.
2. But prior to that it should be very relevant to refer to I.F. Spellburg's (1996, in *Conservation Biology*) suggestions in ecological restoration.

However in simple language issues to be tackled to improve habitat are :

– Fire	– Over exploitation of boulder
– Grazing	– Over grazing
– Flood	– Raising plantation of single species
– Flood	– Aquatic site
– Landslide	– Abandoned mine sites

DESIGN AND IMPLEMENTATION OF HABITAT (ECOLOGICAL) RESTORATION

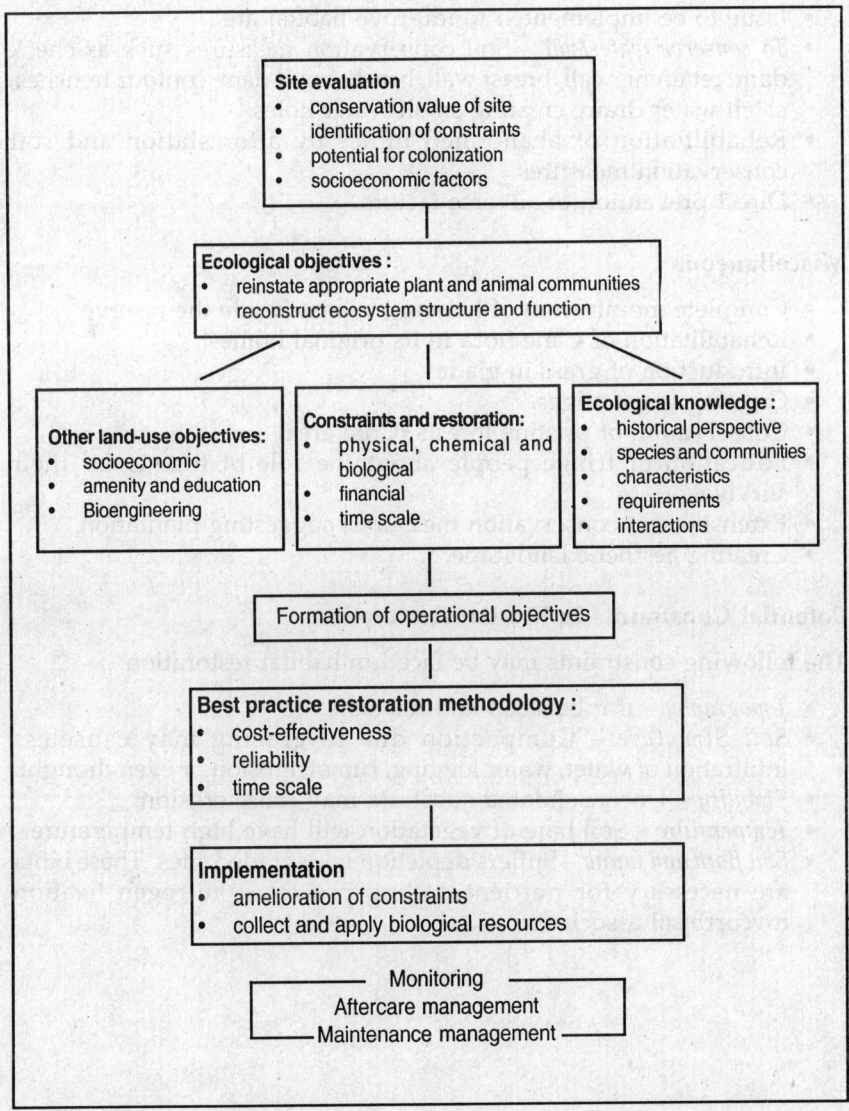

Site evaluation
- conservation value of site
- identification of constraints
- potential for colonization
- socioeconomic factors

Ecological objectives :
- reinstate appropriate plant and animal communities
- reconstruct ecosystem structure and function

Other land-use objectives:
- socioeconomic
- amenity and education
- Bioengineering

Constraints and restoration:
- physical, chemical and biological
- financial
- time scale

Ecological knowledge :
- historical perspective
- species and communities
- characteristics
- requirements
- interactions

Formation of operational objectives

Best practice restoration methodology :
- cost-effectiveness
- reliability
- time scale

Implementation
- amelioration of constraints
- collect and apply biological resources

Monitoring
Aftercare management
Maintenance management

- Plantations are structurally simple, single tree vegetation layer supports few birds.
- Plantations lack den and nest sites since trees are not allowed to get old, rot, and develop hollows
- In plantations food is less diverse and less abundant.

These issues have been analysed in detail in the text and is not being repeated here.

- Issue to be implemented to improve habitat are :
- *To conserve water/soil* – Soil conservation measures such as check dam, retaining wall, breast wall, brush wood dam, contour trenches, catch water drain, creating small water holes.
- Rehabilitation of abandoned mines by afforestation and soil conservation measures.
- Direct prevention of adverse factors.

Miscellaneous

- Complete moratorium of human trespass inside the reserve.
- Rehabilitation of Cane flora in its original homes.
- Introduction of grass in glades.
- Creating forest edges.
- Conservation of pristine forests (Core area)
- Education of fringe people about the role of forests for their survival.
- Extensive soil conservation measures, suggesting plantation.
- Creating aesthetic landscape.

Potential Constraints in Habitat Restoration

The following constraints may be faced in habitat restoration:

- *Topography* – if it be steep and unstable.
- *Soil Structure* – Compaction due to grazing may causeless infiltration of water, water logging, run of, erosion or even drought.
- *Stability* – Unconsolidated substrata may cause erosion.
- *Temperature* – Soil bare of vegetation will have high temperature.
- *Soil flora and fauna* – Suffers depletion in degraded sites. These biota are necessary for nutrient cycling processes, nitrogen fixation mycorrhisal association.

FACETS OF PLANTATION

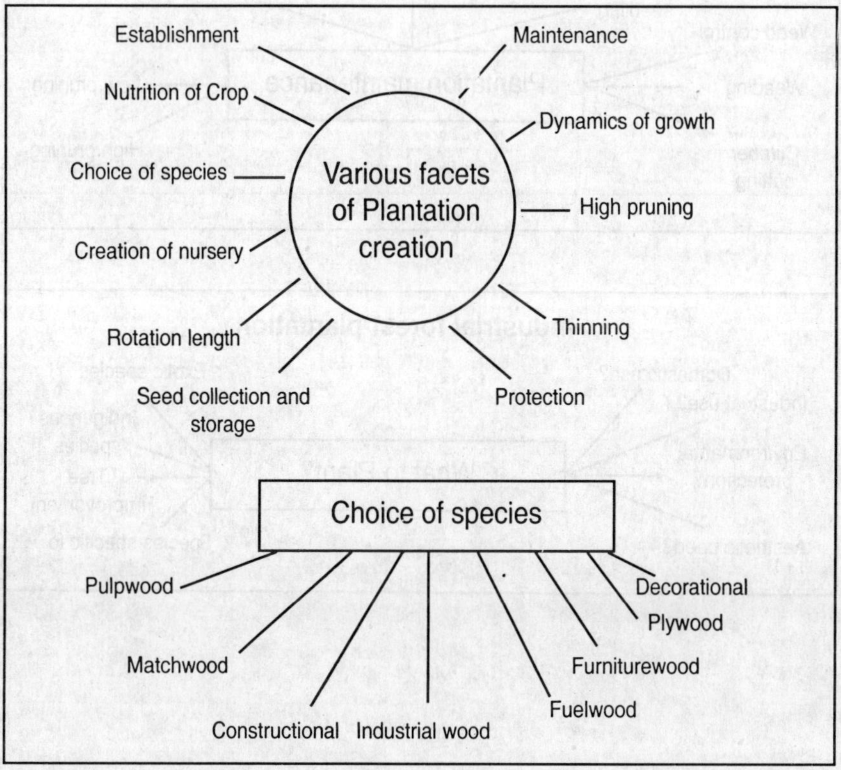

Establishment

Maintenance

Nutrition of Crop

Dynamics of growth

Choice of species — Various facets of Plantation creation

— High pruning

Creation of nursery

Thinning

Rotation length

Protection

Seed collection and storage

Choice of species

Pulpwood

Decorational

Plywood

Matchwood

Furniturewood

Fuelwood

Constructional   Industrial wood

WORKS INVOLVED IN PLANTATION

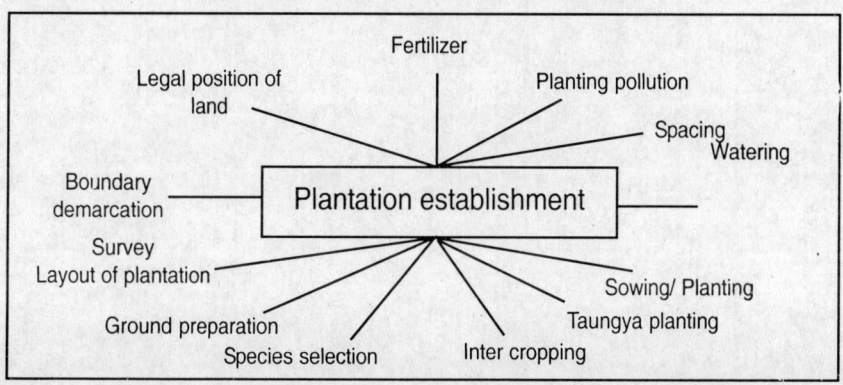

Fertilizer

Legal position of land

Planting pollution

Spacing

Watering

Boundary demarcation — Plantation establishment

Survey

Layout of plantation

Sowing/ Planting

Taungya planting

Ground preparation

Species selection

Inter cropping

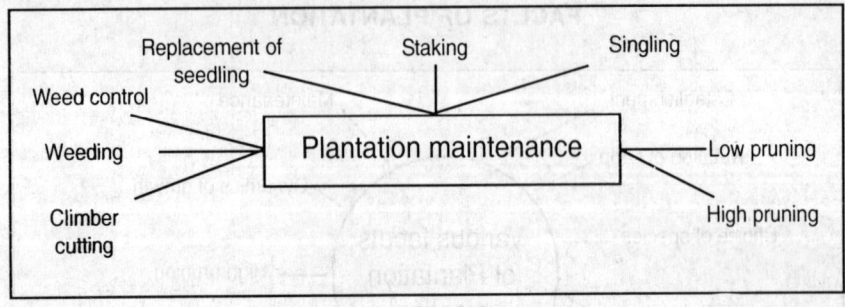

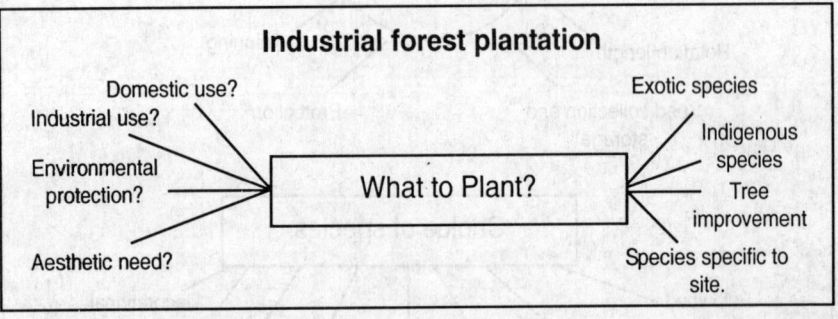

13

Forester's Contribution to Ecodevelopment and Sustainability in Protected Areas
— A Case-study in Moist-deciduous Forest

It is needless to mention the important role of forester's and their contributions towards the efforts for maintenance of ecological balance. Their role have become more important in view of the dwindling forest resources and wildlife due to population explosion. Considering the present state of affairs, the forestry experts have put maximum stress on eco-development and sustainability of the forest resources. In doing so the forest fringe population have been brought under the umbrella of coordinated forest management by creation of eco-development committees (EDC) and forest protection committees (FPC). Forest fringe population are being motivated to become self-sufficient through various schemes and take part in protection of the forest and wildlife. The authors made in-depth study in respect of sustainability and eco-development in Buxa Tiger Reserve and some of their observations have been recorded in this chapter.

13

Forester's Contribution to Ecodevelopment and Sustainability in Protected Areas
— A Case-study in Moist-deciduous Forest

The subjects have very wide perspective and the authors do not discuss the evolutionary growth of the project since it's embryonic stage (1973) as the purpose of the present compilation is different; their intention has been to bring in view the initial success of the project in its first 20 years of life and if it suffered from several derogatory impacts that slowed down the progress of project. At that juncture the concept of Ecodevelopment was born and the management of the project. At that juncture the concept of ecodevelopment was born and the management of the projects got input from a different sources. The case study was made in Buxa Tiger Reserve where biodiversity and sustainability study was made. Various problems were identified and remedy suggested. Also ecodevelopment study was made to throw light on sustainable development.

India ecodevelopment project was designed for five years from October 1996. The project beneficiaries (local people from the PAs who will get benefit from the project and will contribute 25% of village development investments in the form of labour). Four broad objectives were identified which are :

1. Improving PA Management.
2. Village Ecodevelopment.
3. Education/Awareness and Monitoring/Research.
4. Research and impact monitoring.

Improved P.A. Management

Since the early twentieth the foresters felt the necessity of creating plantations of commercially important species by clear felling of forests as natural regeneration of sal and other species were absent. For this the foresters provided settlements of labourers in forest villages for raising of nurseries and artificial regeneration of plantations. The village population structure multiplied several times, cattle, allocated to the villagers also multiplied manifold in population with passage of time. This situation has evolved problems relating to grazing, collection of fuel wood and timber and N.W.F.P., forest fire, encroachment of forest areas, poaching of wildlife, etc. which caused substantial loss of biodiversity of commercially important species. Along with the establishment of forest villages, large number of Tea Estates were established in the fringe areas. The Tea Estate population have also become very much dependent on forests.

Several revenue villages existed in the area with local inhabitants with very low density of population. After partition of country, several thousands people migrated from erstwhile East Pakistan settled in these villages. Afterwards also, many people migrated in the area as a result of ethnic disturbances in the adjoining border of B.T.R. These population also exert considerable pressure on the resource of B.T.R. All these have been discussed in the text.

Need for Baseline Data

Biodiversity conservation is the basic need for sustainable development of various activities in a project area. For this the imperative necessity is to start from a baseline reliable qualitative data, which the B.T.R. did not possess on various issues. This situation prompted the authors to study various biotic and abiotic factors and their impact on the resources and physiognomy of B.T.R. This was necessary to evaluate the degree of loss of biodiversity. Unless these derogatory factors are removed there would be little chance left for the development projects meeting with success. Some of these factors are fire, grazing of cattle, boulder collection, long years of forest management for commercial forestry, indiscriminate collection of N.W.F.P., flood, erosion, theft of forest produce, poaching, man-animal conflict, heavy demographic pressure, etc. All these factors have been brought under the sphere of study.

The result of study of resources, floral and faunal, has been detailed in several checklists which establish the resources identified till today should be considered as index of minimum assemblages of resources.

Methodology of Study

The present project executed in B.T.R. had a multifaceted ecological economic and social edges. The vast forested tract was faced with relentless various biotic and abiotic impacts. A heavy human and cattle population was causing depletion of resources everyday. In such a terrain and a problem ridden tract the issues had to be studied and tackled from various angles and approaches.

The project area was thoroughly visited and the authors made a rapid survey of the project area, held discussions with concerned authorities, field staff, the stakeholders and others. The authors had to go through available reports and documents and be equipped for preparation of this work.

Thereafter, the core personnel of the team supported by suitable number of field investigators and analysts visited the area and interacted with all concerned and an Interim Report was prepared on the basis of the same.

The research methodology for sustainability study was made on individual and group interviewing of various stakeholders, regarding observations and gathering information from various documents. The heart of research system was a series of semistructured interviews with villagers (the beneficiaries and other villagers, local managers), forest staff, Project Officers, NGOs, etc. (stakeholders). Interviews were combined with a variety of rapid diagnostic tools [swoy, pra/rra and other observation techniques] which allowed the villagers to participate.

Search for Relevant Reports and Literature

Initial duty was the collection and understanding of various available reports, literatures and documents, some of which were :–

- Administrative reports by Civil Official of Nineteenth century.
- Various Forestry Report since, 1870.
- Various Working Plans (Management Plans of Buxa Forest Division).
- Various Management Plans of B.T.R. since, 1983.
- Guidelines of G.O.I. for Ecodevelopment.
- Guidelines of G.O.W.B. for implementation of Ecodevelopment programme in and around P.A.
- Village Ecodevelopment plan.
- Information about capacity building and training of staff and community.
- Information on conflict management means and methods.

Hazards of Biotic and Abiotic Factors

Assessment of impacts of these factors were made by visit of:

- Flood affected areas and biodiversity lost.
- Fire affected areas and damage caused to biota.
- Erosion affected areas and the problems.

- Cattle grazing areas and the biodiversity suffered.
- Forests in other areas.

Ecodevelopment Related Study

The following methodology was adopted during the course of study
- Selection of beneficiaries for supply of inputs.
- Attendance in village meeting : Usual participants and issues discussed.
- General idea about income loss due to stoppage of forestry operations in P.A. management and alternative income generation with example, indicating constraints, if any.
- Amenities of staff.
- General feeling of community and grass roots level staff about project and their suggestion for improvement.
- What are the constraints of income generation for most of the people who depend for their livelihood from collection of fuelwood
- Is there any dependency relationship between the poor villagers and outside middlemen for supply of goods either produced by the villagers and / or forest produce in lieu of the money provided to them in advance in critical period?
- What is the level of awareness and motivation for conservation of forests for sustainable living?
- Is there any village fund maintained for village development?
- What are the contributions of the villagers to the project? How is it accounted for?
- What is the general reaction of people for cattle improvement, stall feeding, use HVY varieties of crop, double cropping, irrigation, organic manuring, cottage industries, wood saving stove, etc. The ideas in each one of the improved practices to be discussed to understand their present situation, their constraints and opportunities.
- Ascertain the strength and weaknesses of the community organizations so that necessary suggestion can be put forward for their capacity building.
- Assess method of decision taken by the communities and participation of women and Tribals in the matter.
- Examine the guidelines and gather socioeconomic and ecological information, including cause and effect relationship, needed to improve the guidelines.
- Assess the condition of the communities affected due to establishment of P.A. and the means of alternative livelihood, of those communities and their sustainability.
- Adequacy of funding source, slowing and halting of unsustainable resource use.

- Meeting basic needs of communities.
- Support Joint Forest Management.
- Assess negotiating skill and suggest strengthening the capacity to negotiate with the outsiders like traders, etc.
- What are the criteria for selection of target of ED area?
- The list of NGOs working, what are the criteria for selection.
- What are the various AER for P.A. and villages and villages selected under each?
- Credit system in operation in the villages.
- Marketing system.
- P.A. specific items of works, scope of works, employment generated, source of labours, etc.
- Organizational chart.
- Eco-restoration works.
- Guidelines for management followed and results achieved.
- Collection NWFP, who are the people collecting and its marketing.

Vegetation Analysis in Sustainability and Biodiversity

- Quantitative analysis of vegetation was made in plot laid statistically following accepted ecological methodology to ascertain density, cover, frequency, abundance of flora and compare the results of various areas using diversity index and similarity index.

Faunal Status

- For animal, bird and Ichthyofaunal studies available information were collected and carefully analysed for sustainability study.

Microplan and the Fringe People and Forest Villagers

- Examination of some selected microplan was made over widely selected sites and the concerned stakeholders were interviewed, various projects implemented by them were inspected in the field and their social and economic status, attitude towards forest activities, awareness of the benefits were identified.

Miscellaneous Issues

- Preparation of questionnaire for collecting various data from villages and Tea estates.
- Collection of information from Field Director on smuggling, theft, poaching and court cases, etc.
- Interviewing FPC, EDC and JFM committee members about the success and deficiencies in project activities for the purpose of mitigating their problems.
- Assessment of unemployment situation in the fringe area.

Village Ecodevelopment

Summary

The authors kept in view how to reduce the negative impact of local people on biodiversity, reduce negative impact of P.A. on local people and increase collaboration of local people in conservation efforts.

The aforesaid objectives have been implemented in the following way:

- Microplanning and support for implementation
- Reciprocal commitments, comprising measurable actions to improve conservation and associated investments that foster alternative resource uses and livelihood and
- Special programme in joint forest management, voluntary-relocation and discretionary funds for special needs.

Appraisal of socioeconomic and political environment, characteristics of the communities and their traditional mode of living, history of settlement, population pressure, etc. were carried out to identify constraints and opportunities for eco-development, prior to study of the process of implementation.

The project authority prepared micro plan fairly exhaustively and use several PRA tools for participatory activities to initiate implementation.

The resource base on the forest (NWFP) has not been considered for income generation. One of the important issues of FPC/EDC of BTR is that the members are from various ethnic groups with different mode of living, skill, attitude, entrepreneurship, literacy, background, etc. The issue has important bearing on their motivation apart from income generation support for eco-devlopment, which needs due consideration.

The micro plan provides list of activities with the assessment of costs and also level of investment by the project determined on the basis of Rs. 12,500 per family which has been viewed as individual entitlement while implementation. Some of the micro plan activities failed to link up with conservation.

Several promises have been made in the micro plan for conservation will not have any effect unless these promises are internalized by the FPC/EDC and change their attitude. It is necessary that the community should have livelihood support from the conservation of resource BTR. This will enable them to realise the importance of conservation. This may be in the form of employment generation and or income from WFP. The micro plan does not appear to have taken into consideration of above factors or allied activities. It has made general plan mainly for agriculture development, though it is true that land base income generation activities with intensification and diversification can lead to substantial income generation. But the under special circumstances of BTR other opportunities should have been explored in view of large number of marginal and landless people. The various agricultural activities (cultivation of crops, dairying, fishery) have been taken up FPC/EDC-wise with the investment provided by the project with the idea of developing attitude of togetherness to carry out eco-development. But the field study did not provide enough indication of ownership of the activities by many of the members. The enthusiastic are the leaders of the committee about the activities. Since the activities have been started little more than one year, it is too early to obtain result, but the issue of equity, ethnicity, gender need to emphasised for sustainability of such venture.

Similarly, village committee formed is not functioning satisfactorily. The organizational culture has not yet developed. There is hardly any shared value, belief and behavioural norm. In view of limited capacity of majority member few capable persons have been entrusted with the job of running the organization. There is no system of support for capacity building of communities to run the organization and storing and disseminating information among its members for development of common outlook.

The project has been not developed firm linkage with various sectors (productive and social) to provide technical support to beneficiaries.

The most promising activity is formation of SHG (Self Help Group) of women. The micro credit operation by the group may help them to provide income opportunities and in conservation efforts due to development of group solidarity.

The negative factors of conservation are allurement of earning high income from illegal felling of timber and high degree of unemployment.

From the study transpired that overall socioeconomic and political environment and heavy demographic pressure, which is fast changing, have been offering a great influence on the sustainability of project intervention which may be beyond the limit of BTR to resolve exclusively.

On the contrary, there are several indications of culture of participatory management recorded during the period of study that may have influence on sustainability of eco-development activities.

Provisional sets of recommendations have been provided at the end of the study.

Ecodevelopment in Village Level

Objective of Ecodevelopment

The objective of this component is to reduce negative impact of local people on biodiversity of PA and increase collaboration of communities in conservation efforts, through (a) micro-planning and implementation support; (b) reciprocal commitment comprising measurable actions to improve conservation and associated investments that foster alternative resources uses and livelihood; and special programme in joint forest management, voluntary relocation and discretionary fund for special needs.

Appraisal of Socioeconomic and Political Environment

Background

The people living in forest fringe areas from multi-ethnic groups (Rava, Boro, Garo, Oran, Bhutia, Nepali, Rajbansi and Bengali). The Rajbansi, Bengali and Nepali are dominant population in revenue villages, whereas forest villagers are mostly Rava, Boro, Oran, Nepali and some Garo and Bhutias in the hilly areas.

The tea garden villages have similar ethnic population as forest villages. These various groups of people either live together in a village or live in separate locations in the same village. Among the tribal groups, womenfolk is vary active and the society is matriarchal in case Rava and Garo. Some of the clans, especially the Nepalis are more enterprising than others with respect to saving habit and investment. They also have higher percentage of literacy with strong family bond and provide higher status to women in comparison to Oraon. The clans like Oran, Rava, etc. do not have saving habit but drinking and borrowing habit is common among them. Their womenfolk are most exploited. They work hard in the outdoor to earn for the family and are also responsible for household works. The male-folk are comparatively lazy. The extort hard earned money from their wives and even borrow from moneylender for drinking and gambling. Some of the villagers have a feeling that these forests have been raised by their forefathers, so they have the right on the forest for collection of forest produce.

The development activities, which are being carried out under the project, are taken as government work. Many of the villagers interviewed do not feel the project is for them, they consider it as government work. Actually ownership of the project and sense of belongingness of the communities is lacking and there are gaps in information about project aims and object.

Settlement of Forest Villages

The forest villagers were settled for forestry operations and Forest Departments (FD) provided them land for cultivation (0.2 ha for homestead land, 0.4 ha for permanent cultivation land in hills and 1.0 ha for permanent cultivation land in plains), facility for *taungya* plantation (inter-cultivation of agricultural crops in between forest plantation lines) in lieu of raising forest plantation, wage employment to carry out felling, logging of trees and other forestry activities, permission of maintenance of limited number of cattle, fuel wood for domestic use, and pole for repair of house, permitted to collect edible yams and other NWFP, etc. In course of time with rise in population in villages, forest villagers encroached considerable forest area for expansion of cultivation, and also increased the number of cattle per family.

Traditional Income Opportunities of Forest, Revenge Villages and Fixed Demand Holders

The regular plantation forestry started in Buxa since 1920 due to failure to obtain natural regeneration of sal though experimental plantation started in the later part of nineteenth century. The forest villages were set up to meet the need of labour. First forest village was set up in 1904.

The target of raising plantation increased manifold for rapid conversion of high forest to plantation to create a reserve of high crop since sixties from the development fund available for the purpose. The creation of large hectare teak, sal and miscellaneous plantation through village and departmental *taungya* (growing of agricultural crops departmentally by wage labour in between forest lines) continued till the felling of the area was suspended. This has resulted creation of large areas of plantation forests (33.3% of area).

The increased felling created additional employment opportunity, which was even met from surplus labour of tea garden and revenue villages. The wage employment generated prior to stoppage of clear felling in 1984 was 0.59 million many days. Since early seventies the forest department providing wage for the forest plantation raised in conjunction with agricultural crops (*taungya*) as an outcome of movement by the forest villagers. The tiger project was initiated in Buxa in 1983 and India Ecodevelopment project was launched in September 1996 because of its importance as biodiversity reserve. So in Buxa Tiger Reserve (B.T.R.) nominal felling was confined to the periphery indifferent plantations and blanks for biodiversity conservation in core area as a result of which wage labour employment was reduced to 0.28 million may days in 1991.

Usually revenue villagers were not involved in forestry operations. They were dependent on agricultural and allied activities and wage employment. These villagers had no legal entry in forest areas prior to formation of FPC/EDC. They however, used to collect forest produce by permit on nominal payment.

The fixed demand holders were involved in handling and trading of timber, firewood, service, petty business, mining, etc. prior to formation of B.T.R. They also maintain large herd of cattle, which graze in the forest. These people are affected due to stoppage of felling, mining, etc. The literacy percentage of these people is higher than other groups of villagers; naturally the percentage of educated unemployed is also higher.

Due to drastic reduction of wage employment after constitution of B.T.R., the villagers have become more dependent on forest. They took recourse to diversified strategies of collecting and selling firewood, timber, collection of edible tubers, leaves, etc. for consumption and wage labour job whenever available. Thus pressure for foraging forest produce increased substantially. The occurrence of Dioscorea (an edible yam), cane and medicinal plants like Rawolfia, has decreased considerably (the earlier study recorded at least one yam of Dioscorea per acre and now the species has become rare as recorded in the present study).

Similarly the middle storey forest species have disappeared from many areas, may be due to increase collection of fuel wood noted during the study.

State of Illegal Operation of Timber at Present

It has been gathered during study from local people that there is log of allurement given by the unscrupulous businessmen to the villagers for felling and transporting high value teak trees from plantation area. (A 20-year old teak trees produce 8-10 cft of wood. The market value of the same is minimum Rs. 3,000. The entire produce can easily by carried by a van rickshaw). Many people interviewed are of opinion that forest villagers are actively taking part in illegal activity. Large number of unemployed and underemployed youth in fringe areas and tea gardens pose a major problem as corroborated by the fellow villagers. During monsoon period, most of the B.T.R. becomes inaccessible and river rafting of illegal timber becomes common feature. As a result species like *khair* and *sisso* have disappeared from the forest block adjoining Rydak and Sankos river. Some people especially from fixed demand holdings mentioned that there are illegal migrants in the forest villages who are engineering the illegal operation. The forest villagers do not accept their involvement, but social acceptance to people involved in illegal felling is not denied. The easy way of earning by pilfering high value trees may have far reaching consequence on the habit and attitude of the villagers. This may make them completely disinterested for earning by hard labour for income generation.

Village Institution

Each forest village had a village headman nominated by the villagers who was primarily responsible for allotment of areas among villagers for raising plantation prior to formation of FPC/EDC. He was the contact person for mobilizing the villagers for taking up various activities in the forest areas. The revenue village has *Panchayat* institution.

Present Supply and Demand of Timber and Fuel Wood

There are 279 recognized fire wood dealers, 26 saw mills, 45 cane processing units, 4 veneer units. Due to suspension of felling, supply of timber and fuel wood decreased from government source drastically. The various persons involved in timber and fire wood trade were affected. Many saw mills were closed down. The fuel wood dependent population adjoining forest area were 34 villages and 4 fixed demand holdings (population 17,750) 46 Khas villages (population 97,165), town and tea garden (34 tea gardens adjoining reserve with population of 1,72,268) (all 1991 census figures). On the basis of figures available for 15 tea gardens out of 34 located fringe area (M.P.O. 2000) estimated tea garden extracts about 25,000 tons fuel wood. At present tea gardens provide Rs. 100 per labour family allowance for fuel in view of non-availability of allotment from B.T.R. Recently the project has offered some stacks of fuel wood to sea gardens that they have not lifted because of availability of cheap fuel wood. The fire wood is available in the market in the town as gathered during the study. Large number of villagers especially women irrespective of villages gather fuel and sell in the market.

The forest department sale depot lots mostly comprising timber collected from the forest which has been left by the timber thieves and also seized produce are mostly of teak and sal and produce obtained from limited felling area. But in most cases they do not get good price. The average market rate of teak is less than government scheduled rate mentioned by the forest corporation dealer at Alipur duar. A school teacher constructing a house when interviewed about source of teak timber being used by him said he gets home delivery at cheap rate. In management plan mentioned that timber from forest reserve is exported not only to Kolkata but also to Bangladesh and some areas of North India.

Population of Human and Cattle

In course of years the population of forest villages increased manifold. To cite as an instance a forest village set up with 32-household has a present 214 households. The population of project district Jalpaiguri increased by 25.95% between 1981 to 1991 report (1991 census figure). The total population of forest fringe areas of revenge villages is 97,165 and forest villages including fixed demand holding of B.T.R. is about 17,750 (1991 census figure).

The tea gardens are outside of the protection committee. The population figure for tea gardens is 1,72,268.

Similarly there has been tremendous increase of cattle population in forest of khas and tea garden together is 2,00,000. Intensity of grazing is 2.5 cattle per ha.

Employment Situation

Large number of unskilled labourers residing in forest and fixed demand holding villages who were primarily dependent on forests have lost their job opportunities due to stoppage of felling and regeneration. The large number of educated unemployed involved in secondary and tertiary sector activities have lost their scope of self and wage employment. The employment situation in tea gardens is also not satisfactory.

Road and Railway Network

Most of the forest areas and 70 per cent villages are well connected with roads. The national highway also passes through the forest area. Various goods and transported by large number of trucks (Punjab body) to various places in North East India pass through the forests of B.T.R. On return many such trucks transport illegal timber from these forests. The railway line passes through the forest area. There are several rail heads in and around forest area which facilitate illegal smuggling of timber. Several accidents of wildlife take place because of railway line. The B.T.R. staff seized several trucks, van rickshaws for carrying illegal timber found in various range offices.

Poaching

There are several incidence of poaching in B.T.R. during 1979-97, out 27 elephant poached in West Bengal, 15 were from this reserve.

Problem of Tea Industry

The Indian Tea Industry is facing stiff challenge at home from tea imported from Sri Lanka, Bangladesh and Kenya thus limiting prospect of its expansion. In fact some of the tea garden people fear that increasing foreign competition along with low level of productivity of the tea garden of the region may lead to closure of sizable number of garden. If that happens the B.T.R. may face biggest challenge for its existence because pressure of labours of closed garden.

Agriculture

Agriculture in forest villages is carried on by traditional method. They grow local variety of *Aman* paddy in low land. They usually do not grow winter crop in most cases. They hardly use any manure or fertilizer for cultivation. Irrigation facility is limited in forest villages. The cultivated land is also subjected to flooding and erosion located near the streams. The depredation by wild elephant is a disincentive for growing food crops. In the past they used to grow jute when it had good market. Forest villages are not within *panchayat* and agricultural extension service facility is not available. Due to settlement in government forest land they are not eligible for bank loan.

Most of the villagers have homestead land, which is not used judiciously for growing fruits and vegetables by majority. Some enterprising villagers grow various fruits and vegetables. Again threat of wild elephant is disincentive for cultivation of banana and bamboo.

Villagers maintain large number of unproductive cattle. These cattle are maintained by grazing in forest land. There is hardly any income from cattle. There is no facility for scientific feeding, breeding and maintenance. Many villagers maintain pigs, chicken and duck.

Health, Education Supply of Drinking Water and Communication Facility

In general health and education facility is not satisfactory. Some of the forest villages have primary schools. Drinking water facility is available in many villages and some action has been taken to create facility in eco-development project. Similarly some village roads are constructed from the fund support of the project.

Political Environment

There are large number of political parties and trade unions in the area and in any given time these parties are indulging in a battle of increasing influence among local population. It would be suicidal for a political party to vociferously oppose deforestation. In fact some political party claimed having lost *panchayat* election because of having taken a principle regarding forest protection in the conflicts between forest and local people.

There are 13 staff unions in B.T.R. They are involved with various issues for the benefit of members. The Project Director and concerned staff are engaged in considerable time in meeting and discussions with the unions. As a result a several unions, large number of staff are engaged in union activities. It is the great experience in B.T.R that most of the cases of implication of staff in illegal activities takes a political turn.

As a result the whole process of detection and punishment ends up with a fiasco. This has a demoralizing effect on sincere and honest staff.

Findings on Ecodevelopment Activities

Attitude of Beneficiaries

Attitude of beneficiaries towards eco-development activities are mixed. The women folk in general is very much interested for the income generation activities and very positive in their thoughts and doing, while many men feel that eco-development is FD activities. Further many villagers of forest village feel that these forest have been raised by their forefathers, so it is their property. So, they have every right to fell and sell the tree. The unscrupulous people living outside the forest area allures the villagers for felling and carrying the valuable teak on payment of handsome amount.

The growing desire of modern living and purchase of consumer goods were noted among many villagers (some of the forest villagers possess TV). The accessibility of quick money and desire of consumer goods may be responsible for not been very enthusiastic about eco-development activities to many villagers especially located in accessible areas. But the story is different in the villages in the remote areas.

Preparation of Micro Plan

The project prepared 52 micro plans covering 8,431 families including 4,296 tribal, 2,250 landless and 500 women. There was delay in preparation and implementation of micro plan due to limited capacity of project staff and problem of obtaining contribution from the beneficiaries.

Micro plan has been prepared fairly exhaustively and has used several PRA tools for planning. But plan lacks in information about forest dependent vulnerable household and women headed household. So, no action has been taken to form subgroups of these vulnerable population to reflect their concerns facilitate to design income generation activities based on the availability of resources and their capability. The selection of investments have been made more on scheme centred than on capability of people. The micro plan also does not provide any information regarding type of resources available in the area and skill potential of the communities, identification of training needs of the communities for capacity building to plan and implement appropriate eco-friendly income generation opportunities. Detailed analysis of different options in the micro-plan area should have been made available to explain the villagers for informed decision making. Similarly in case of group activity (like that of winter cropping, fishery, dairy, etc.) details of land/asset ownership, income flows and distribution, and mechanism for sustaining the operation should have formed integral part of micro plan. Since multidisciplinary team was engaged for planning, it would have been appropriate to indicate training need for capacity building of forest staff, NGOs and communities for implementation. On the contrary, the micro plan provides a list of activities with assessment of costs and also level of investment by the project determined on basis of Rs. 12,500 per family which has been viewed as an individual entitlement while implementation. Some of the micro plans activities also have failed to establish linkage with conservation. There is hardly any information regarding feasibility of sustainable collection, processing and marketing of NWFP for income generation. In general, planning has a bias for rural development that eco-development.

Micro plan indicates FPC/EDC members made several commitments for protection and conservation of forests which are very ideal but how the ordinary members have internalized these commitments is doubtful as could be ascertained from several members interviewed because of their different expectation due to different background, ethnicity and literacy percentage. Besides overall environment, because of large scale illicit operation, poaching, grazing, etc. are also not congenial for conservation as appraised during field study. However, the committees have contributed 25 percent of grant and they have deposited the money in the bank as per commitment. No rules and regulations have been formed about utilisation of the fund in future.

Forest departments in their turn have fulfilled many of their commitments. The mutual trust between villagers and forest staff have not yet developed in may cases.

Institutional Arrangements

The earlier village institution for forest village operation is nonexistent. Instead each village has an executive committee comprising the six to eleven elected representatives of village, *pradhan* or any member of *gram panchayat*, a member nominated by *sabhapati* and beat officer as member convenor as per G.O. for constitution of EDC/FPC. But the village committee formed is not functioning satisfactorily. The organizational culture has yet to develop. There is hardly any shared value, beliefs and behavioural norms. In view of limited capacity of communities in general, one or two capable person has been entrusted with the job of running the organisation, maintenance of records and running the committee. There is no system of support and encouragement of bottom up empowerment by the organization. There is no mechanism for gathering and disseminating information by the organization among the members for effective functioning. The general body has been formed where all villagers meet once a year to discuss various issues and appraise eco-development benefits, etc. In some villages resolution for action is taken but follow up action is hardly pursued. The large membership in EDC/FOC makes effective functioning of the committee difficult.

Since 1998 the forest villagers have been included in *panchayat* and many of the forest villagers are members.

Implementation Process

Capacity building of project staff, NGO and community.

The capacity building exercise consists of team building, PA management training and skill development training, and course on veterinary care and study tour, besides training for preparation of micro plan. In this training and study tour project staff, NGO and communities of both genders were included.

Project Staff

The project staffs are technical trained in forestry but by and large have limited exposure to participatory management and institutional development. Of course there are some very imaginative and innovative Range officers who have done excellent job especially in West Division. The project needs multidisciplinary input, which is not available within the project. Besides linkage with other departments are practically absent in most cases.

The senior project staff spends substantial proportion of time in various meetings and discussions with various bodies and staff unions. The field staffs are largely engaged in eco-development activities because of new type of job. They have limited time available for protection.

NGO

The local NGOs were also engaged for implementation. Many of them moved into the area because they know forest well. They are more in the nature of local clubs. They have limited knowledge in the area of participatory development and little technical knowledge. They do not have any previous experience or training in institutional development which is key to initiate participatory development process. They have been subcontracted for the delivery of the project inputs. The midterm review mission of the project has recorded in their report that NGO: assisting FD in undertaking the micro planning have generally poo understanding of Ecodevelopment and also indicated even the bigge NGOs like IBRAD have focused on traditional rural developmen activities rather than activities linked to conservation needs.

Gender

The participation of women in the preparation of micro plan preparation process has been uneven. The special importance of woman role in the tribal society for income generation has not been given for preparation of micro loan. In general burden of work of women is increasing due to provision of additional livestock provided by the project in most of the poor household. They even expressed to the consultants about their hard work for collection of fuel wood and selling to earn living. The women in many villages are interested in loom weaving. They prefer to take up weaving and or some income generation opportunity in place of fuel wood collection. They have also started using fuel wood saving stove in many villages observed during the trip. Women in some committees are playing crucial role in community patrolling.

567 women have formed 32 self-help group (SHG) and made a total saving of Rs. 70,691. They need support for accessing institutional finance, skill training, ideas for income generation activities, managing revolving fund, etc. Some SHG have been provided training CARE on the concept of thrift, credit and also on health and nutrition.

Though initially the project facilitated to form SHG but at present women themselves are coming together. In Panialguri village, women formed SHG to counter alcoholism in the village men who were involved in illicit felling for income. These groups are very active in patrolling of forests and have countered timber smuggling, which has been reported to the press.

Implementation of Activities

The implementation activities were slower and in some cases delayed due to several reasons (a) the friction between FPC/EDC members and FD due to detection of illegal felling and poaching, (b) involvement of some FPC/EDC members in illegal activities after commencement of micro-plan operation, (c) lack of reciprocal commitment, (d) lack of experience in implementation in such diversified activities (e) delay in micro planning (only a few micro plans were prepared till 1-6-1999).

In the initial stage of implementation, the project selected beneficiaries based on their ability to contribute in Village Development Fund (VDF) which is 25% of project investment and provided them input for their income generations as a result poor forest dependent people could not get any benefit. Later on poor people were provided wage employment to contribute for VDF and obtain fund for income generation. In Turturikhand (an area of Bhabar tract) 204 villagers of the area contributed voluntary labour to provide VDF to obtain fund for road construction and deep tubewell.

The income generation activities are being implemented on individual beneficiary basis in eastern side of project area and on community basis in western side. In case community basis of implementation as noted Nimati Domohini area the community has obtained total fund based on number of families in EDC from the project and contributed 25% VDF, but the contribution is not from households. The Range officer informed that those who could not contribute would gradually meet their dues to those who invested on their behalf. The entire fund at a time was necessary to start the micro plan activities and accordingly this policy has been adopted as mentioned by the Range officer. It was further recorded that community activities are carried out by group of people of the village on wage employment and the income obtained from the activity is retained by a member of executive committee to meet the running expenditure. But VDF is retained in the bank. The system of distribution of dividends from the investment has not been developed. The observation of some of the micro plan implementation are as follows:

Agriculture

Several crops like wheat, potato, vegetables, boro paddy have been introduced after harvesting of traditional *aman* rice. The infrastructure for irrigation has been created in some villages for double cropping. But the success of cultivation of second crop is not uniform. Due to delay in planting, inappropriate varieties, disease and pest or problem of marketing resulted poor return. In certain villages, beneficiaries have identified some promising crop like groundnut, squash, potato and wheat for their area. However no attempt is made there to improve yield and income from *aman* paddy through introduction of improved recommended varieties and packages of practices. The linkage with agriculture extension and marketing is very poor rather nonexistent. In absence of availability of cost and benefit figure it was not possible to analyse cost and benefit. In Checo EDC, an agricultural assistant engaged in contract that is providing technical input and arranging inputs, which has given good result.

Dairy

Improved breeds of cattle have been introduced in some villages individually in eastern division and by community in western division. The community employed few person in the member of executive committee, makes the running expenditure. The cattle have been insured. But feeding and maintenance of animals did not appear to be on the scientific line. The local Range officer has facilitated to establish linkage with chilling plant for marketing but there is no appropriate arrangement for veterinary care.

Poultry Farming

The poultry birds are maintained individually as well as by committees. The results are mixed. There are reports of loss of animals due to disease and problem of marketing of broiler chicken.

Fishery

The fishery is practised in some abandoned pond; and some new ponds dug for the purpose. There is good scope for pisciculture provided scientific practices are taken up. The area imports fish from outsides as there is good market and this will not be affected by wild life.

Weaving

In some villages, women group initiated weaving as micro plan activity. They have been trained and weaving looms, threads have been provided by the project. The trainer supervising the operation has taken the responsibility of marketing. The participants were found highly motivated and they expressed that they would prefer to take up this activity than collecting and selling firewood, which is very arduous.

Linkages for Technical Input

The project has not developed proper linkage with the various sectors for providing technical inputs to the beneficiaries on regular basis. Whatever has been seen is the personal effort of the field staff.

Threats

Forest assigned to FPC/EDC are commercially very valuable. There is network of road, ready market and allurement by the people with commercial interest in one hand. On the other large majority of members are finding difficulty to eke out a living due to non-availability of alternative income generation opportunities. Such conditions pose a threat for community protection for valuable resource.

Tea, timber, transport and tourism are the four major industries of North Bengal. With the stoppage of forestry operations, employment generation in the sector has reduced, the tea industry is in the cross road due to import of cheap tea from tea growing areas abroad. So, problem of unemployment may be the real threat for conservation of forests is foreseeable future.

Growing pressure of grazing, severe soil erosion and flooding, invasion of gregarious exotic weeds in the habitat are also threat to conservation of biodiversity.

Some Positive Indications

Though the project is in its fifth year but due to several initial teething troubles for implementation of this new type of project, implementation started in full swing from little more than one year ago as ascertained from the project staff. The following positive indications have been obtained.

In Panialguri village women group countered alcoholism among men who were involved in illegal felling. The group is engaged in patrolling forest and preventing illegal smuggling of timber.

In many villages, due to setting up of irrigation infrastructure, the intensity of cropping has increased. This is creating extra income to the owner and job opportunities for wage labour.

A beginning has been made to organize forest users to systematize cutting and collection of fuelwood once in a week.

The joint patrolling is being carried out in many areas with success.

There is general awareness for conservation among people. Many of FPC/EDC members participated in procession and deputed to BDO and Police for protection. Some FPC/EDC organized afforestation and taking preventing step against grazing. The Range officer, Damanpur East Range is revisiting micro plan based on resource available and the skill potential of the community.

The Pana Range rood record of the activities are maintained.

There is a tendency found among the villagers of Turturikhand to work together to solve the problem of water scarcity. They are providing voluntary labour to earn project grant to sink deep tubewell.

Linkage has been established in Nemat Domhoni with Panchayat and with line departments.

Local Panchayats of Nimati Range have come forward to extend help to FPC/EDC in their fund support in view of good work done by them.

Gradually many forestry staff are functioning well in their new role as facilitator of human development from regulator of forestry activities.

There is thinking among some of the FPC/EDC Range Officers for grouping of all the FPC/EDC together by location as confederation of FPC/EDC to facilitate institution building, seeking resources from local governments or the District Administration and periodically reviewing the working of FPC/EDCs members to develop rules for utilization of Village Development Fund noted during visit of project sites.

A newsletter in English and Bengali is being regularly published by the BTR authority for dissemination of project information. Local newspapers cover success stories and also illegal operation in project area.

Nature Education and Awareness

Objectives

The objectives of this component is to – develop a more effective and extensive support for conservation and eco-development, through

  (i) environmental education and awareness, improved visitor management and ecotourism and
  (ii) Impact monitoring and research.

It is aimed that management related research and monitoring must be geared up to reduce progressively the extent and degree of uncertainty of which management decisions and management strategies are based.

However, the long term plan is to achieve better understanding B.T.R. ecosystem, functional relationship among the biotic communities and impact of anthropogenic pressure on natural systems.

But, these are not to be easily achieved due to lack of infrastructural facilities, lack of coordination amongst different research agencies and park authorities, lack of financial support and absence of research staff and laboratory.

The Supervision Mission (November-2000) observes that the State Government is already considering the strategy prepared by W.W.F. and is implementing some activities, in collaboration with competent local NGOs. B.T.R. has no education officer and as such they deploy regular protection staff for education and awareness activities.

Nature camps are being organized by to local clubs. B.T.R. has a Nature Interpretation Centre and an Orchid House at Rajabhatkhawa.

Awareness Generation

From the field investigation it was ascertained that the people were conscious about the benefits of forests and the punishment to be faced in case of illegal activities committed. Yet there is hardly any salutary effect on these illegal activities. Even the members of (some sections) FPC and EDC committees were involved in pilferage of timber and other resources, either on their own, or mixing with organized gangs of pilferers.

The huge population of tea gardens and their cattle population cause irreparable damage to PA; no amount of meeting with T.G. officials, who are verbally sympathetic, born result to take active interest in controlling grazing or to contain the cattle either within jurisdiction.

So to combat these threats a thorough change in strategy and strong sincere political will is necessary. Otherwise forests cannot be saved from utter depletion.

However B.T.R. authorities have organized the following :

- Nature awareness camps, organized by Alipurduar Nature Club, Nandadevi Foundation and Mountaineers' and Rovers.
- Through Nature Interpretation Centre and Rajabhatkhawa.
- Organizing Seminars, Workshops, Group Discussions, Film shows, Celebration of Aranya Saptaha and Wildlife Week.
- Excursions.
- Sit and draw painting competition.

Environmental Education Awareness

Implementation Indicators

- National level ecosystem
- Environmental education
- Environmental awareness guidelines
- Ecotourism strategy
- Investments made on amenities and education
- Inventory of local conservation efforts
- Impact monitoring guidelines
- Impact monitoring strategy
- Impact monitoring information
- Research project review panel
- Research project design
- Number of research projects formulated
- Number of institutions contacted for carrying out research projects
- Number of individuals carrying out research projects.

Project Activities

- Develop environmental education
- Training
- Workshop
- Equipment
- Expert advice
- Conduct impact monitoring
- Develop strategies for environmental education
- Ecotourism
- Impact environmental education
- Develop impact monitoring guidelines
- Monitor impacts
- Research for biodiversity conservation

Although awareness activities and implementation indications have been identified in the subject as given below, the present activities are still with initial stage.

Sustainability-linkages to Regional Planning

World Bank Missions in their inspection have put very much emphasis to build trust in the relationship between Park Staff and villagers. Trust building through eco-development activities will need quite sometime. People feel it is FD's obligation to resolve all their problems. It will also take adequate time to mobilize other agencies, private sector and NGOs in carrying out eco-development types of activities in coordination with park staff. Involvement of gram panchayat may be necessary more intimately which will be contacted frequently to offer help for various coordination.

Education and Awareness

Training is an essential and integral requisite to increase the managerial capability, capacity building and technical skill of the staff.

Protected area planning and management is a highly technical science that brings diverse disciplines, i.e. ecology, forestry, wildlife management, hydrology, social science and personnel management, etc. to one platform.

Training has be arranged to staff at various levels (forest rangers, foresters, forest guards, *vana-shramiks*, *mahuts* and grass cutters, etc.). Advanced management training is required to be imparted to the administrative officials to enable better inspection, coordination and innovation of project related activities.

Experienced forest officers, legal experts, public prosecutors, police personnel, research scholars, veterinary surgeons, ecologists and sociologists would be the resource persons.

Name of the course	Duration	Eligibility
1. Postgraduate diploma course in wildlife management at W.I.I.	9 months	DCF and ACF
2. Certificate course in wildlife management at W.I.I., Dehradun	3 months	Forest Ranger
3. Wildlife management training in State Training School (B.F. School) as prescribed by W.I.I.	3 months	DR/FR and F.G
4. Wireless operation and weapon training at Police Academy	—	DR/FR. and F.G
5. Tourism management reception, interpretation and environmental education C.E.E., Ahmedabad	—	ACF and FR
6. Wildlife health, chemical immobilisation, application of power fencing, etc. at W.I.I., Dehradun	—	
7. Capsule courses in wildlife	7-15 days	CCF and CF
8. Remote sensing at IIRS at Dehradun	10 months	DCF and ACF

Training Courses

Wildlife Institute of India (W.I.I.), Dehradun organize various training courses under Government of India assistance or in collaboration with international organisations. Following are the training courses scheduled for a national level.

Some of the important facets of training are:

• Survey and demarcation.
• Tourism management.
• Joint Forest Management and interaction with fringe people.
• Microplanning.
• Wildlife habitat management.
• Monitoring and census of faunal population.
• Tranquilisation of wildlife.
• Veterinary care of animals.
• Cattle improvement, Castration and immunization.
• Nursery and plantation techniques.
• Erection and maintenance of energized fencing.

Other training requirements and last performed by the Park are:

National Park and Eco-Tourism (Study tours) : Study tours were undertaken by MIC, forests, the Chief Wildlife Warden and the Deputy Field Director, BTR (E) in Kenya and Tanzania to acquire first hand knowledge in National Park Management and ecotourism.

Community Forest Management Study (Imparted under W.W.F. Tiger Conservation Programme) : Training on Wildlife Law Enforcement was imparted to the Field Director down to senior and junior field staff under W.W.F. Tiger Conservation Programme.

Elephant census monitoring technique : Training on this subject was imparted to all category of field staff, comprising of senior and junior field staff and *mahuts*.

Tiger census : Tiger census training was imparted to all category of staff (field), comprising of senior and junior field staff and *mahut*.

Census of Herbivores : Census training on identifying and counting of herbivores was imparted to all category of staff (field) including *mahut*.

Staff study : Tour regarding Park Management on I.E.D.P. Works : I.A.C.F., 2 DR/Fr and 4 Fgs were given training on Park Management on I.E.D.P. Works, through study tours.

Rifle shooting training : Forest guards were trained in Rifle shooting under the disposal of Border Security Force, Coochbehar.

Training on chemical tranquilisation : Practical training was imparted to the forest guards on chemical tranquilisation of animals.

Training on care and maintenance of RT System : 24 forests guards were trained in operation, care and maintenance of radio transmission system.

Training on Malaria control : Training on Malaria control was imparted to all junior field staff and F.P.C and E.D.C. members.

Refreshers' Course on Forestry and Wildlife Management : 5 (Five) days' Refreshers' course training on forestry and wildlife management was imparted to 4 batches of 40 heads each of *Banasramiks* and C.D.L.S.

Training on Cattle improvement, Artificial Insemination and Castration : Sixty heads of JPC and EDC members were trained on cattle improvement, artificial insemination and castration.

N.B. : In spite of substantial efforts, rendered for capacity building of the staff and the people concerned in respect of protection of forests and P.A. management, the protection aspect seems to be virtually ineffective. Through interaction with P.A. authorities, FPC and EDC members, NGOs, Timber merchants and other stakeholders including a cross section of the local public from all walks of life it was apparent that the illicit felling of trees, teak on a mass scale and some sal in particular, was rampant in B.T.R. Almost all the depots of the Ranges and Beats are supersaturated with huge quantity of seized logs. Accordingly it can very well be inferred that illicit felling and removal of trees from the P.A. has intensified manifold causing grave concern for maintenance of ecosystem favourable for sustenance of the wildlife population. The habitat is being severely disturbed. As per prevailing circumstances it has also been noted that the bids offered in auction of seized produce from Govt. depots are very low. On investigation it was also observed that there is a sustained decline in sale of teak and sal timber from the authorised dealers of W.B.F.D.C. Ltd. Local investigation revealed that well-off families are getting supply at door step of Wooden furniture made of Teak and Sissoo at exceptionally low rate. All these can very well be considered as negative indicators in respect of protection front of P.A. management, which is considered to be the prime objective of management.

Research and Impact Monitoring

General

During 115 years of Forestry Management and later 18 years of Project Tiger Management, enough laboratory based research data were not generated. The foresters made very useful field-based observations on soil, plants, animals, their ecology and botany.

Since the creation of Project Tiger efforts have been made to involve scientists from reputed Institutions, University teachers, NGOs and NGIs in forest based research.

Research Advisory Committee (RAC) has been proposed to review various proposals and providing them a shape/direction. The RAC was to review, guide and approve research project proposal, with fixation of priorities.

Lack of Facilities

But there has been lack infrastructural facilities like research laboratory and research units. Project authorities could not build up coordination among recognized Institutes and Universities, nor there is any financial support for research and monitoring, Adequate research base literatures have not been collected. Proper data storage and retrieval system is yet to be developed.

Field Research Stations

At Rajabhatkhawa there are some collection of veterinary tools and equipment. Besides there is adequate stock of inject guns and tranquilliser drugs. So far procurement of computers, microscopes, weighing machines, desiccator autoclave, hot air oven, refrigerators or reagents, etc. has not yet been taken up seriously.

Research Officer

No post of Research Officer has yet been sanctioned. The B.T.R. officials are well conscious about establishing a research centre with adequate financial support.

It has been proposed the B.T.R. will established linkages with Research Institutes, Universities, NGOs, etc. to carry out research in biological, social and ecological issues where the Wildlife Institute of India (WIL), Dehradun will play a vital role.

It was also proposed that research issues would have assistance from consultant, research organization, institutes, universities, individual researchers and competent NGOs.

Funding of Research Activities

Indian Ecodevelopment Project is the funding agency. But, collaborative work with other recognized institutions like ICFRE, UGC, CSIR, BNHS, ZSI, BSI and other universities was contemplated. It was desired that IUCN, WWF and other national and international agencies would provide fund for research.

So far no data storage system has been developed, nor there are adequate stock of scientific literatures and journals.

Research Priority Matrix Proposed

Biological

- Impact of cattle grazing..
- Population-habitat viability trade-problems and mitigating measures.
- Poaching and illegal trade-problems and mitigating measures.
- Feeding behaviour of ungulates.
- Study of Tiger niche.
- Study on enumeration techniques of fauna.
- Preparation of flora of B.T.R.
- Distribution of Lesser Cabs.
- Inventory status and distribution of arboreal mammals and other lesser cats.
- Wildlife health and diseases in B.T.R.

Ecological

- Environmental impact assessment of boulder extraction.
- Identification of fragile ecosystem and habitat / niches for conservation.
- Effect of fire influence on flora and fauna.
- Impact of habitat changes on population of Wildlife.
- Habitat suitability study for reintroduction of wild buffalo.
- Effect in grazing in forests by domesticated cattle population.
- Effect of flood on the entire ecosystem.

Socio-Economic

- N.T.F.P. and its impact on village economy.
- Impact assessment of dependence of local people on B.T.R.
- Prospects and strategy for Ecotourism.
- Factors affecting / choice of management regime of BTR.
- Local factors affecting capacity to organize and manage.

Evaluation of Research Output

The BTR authorities contemplates to establish a review committee to evaluate the output of research once in six months! The final report is to be placed before an Advisory Committee.

Research based data on scientific investigations based on statistical and quantitative findings are inadequate BTR. Those available are less assertive than some observations made by the forest officers. No outside institutions or agencies have carried out intensive field-based research either spatial or vertical structures of forests.

Several individuals and agencies given consultancies. It was gathered that few were completed and those reports too were not available with Field Director. The Management Plan Officer in this compilation of Management-cum-Working Plans of BTR has quoted the works of Prof. D. Roy Choudhury on entomofauna and Sri A.B. Chaudhuri of flora.

Topics of Research Intended to be Carried Out

- Man-animal conflict relating to elephant and elephant corridor.
- Impact of cattle grazing.
- Non-wood forest produce and village economy.
- Population-Habitat viability analysis for tiger.
- Environmental Impact Assessment (EIA) of boulder extraction.
- Identification of fragile ecosystem and its conservation.
- Distribution of avifauna in different habitats.
- Status and distribution of cane.
- Poaching and illegal trade – problems and mitigation measures.
- Inventory, status and distribution of abrboreal mammals other than lesser cat.
- Study on prospects and strategy for ecotourism.
- Study on financial sustainability of village institutions (FPC/EDC).
- Population dynamics and feeding behaviour of ungulates.
- Study on flora and monitoring the changes infloristic composition.
- Impact of habitat changes on Wildlife population.
- Study of carrying capacity of the park with respect to important species.
- Study of Wildlife diseases.
- Habitat suitability study for reintroduction of wild buffalo.
- Study on river systems of BTR.
- Study on endangered species of BTR.

Researchers have been desired primarily be conservation oriented; to some extent it should also be management oriented. BTR authorities have fixe some priorities. Research consultancy and study should contribute to baseline information. The researchers should involve local and tribal communities in their research efforts.

The Field Directors is empowered to allocate fund up to Rs. 1 (one) lakh for a project, which is to encourage staff and students of local institutions to undertake research, relevant to Park management.

All research findings from surveys, monitoring and management-oriented research should feed back into Park Management strategies and provide input to materials developed and disseminated as part of education and communication programme.

Forestry Research and Technology Development

Forestry needs research support to improve productivity, reduce losses and wastage, maximise utilization to improve quality and value of plantations for sustainable management of forest resources improved conservation of genetic resources and wildlife, development of N.W.F.P.s and diversification of plantations.

Research studies are also necessary for carrying out preservation and improvement of biodiversity and sustainability and inventory, etc. for providing basic technical data.

Research input is essential for producing and supplying improved and certified seeds, for in-situ and ex-situ conservation of genetic resources. Research input is essentially vital works on.

- Agro-forestry
- Watershed management
- High yield plantations
- Forest management
- Genetic resource conservation
- Forest economics

Research Strategies in Substations

Research substations have been proposed to be established at Damanpur, Jainty and Kumargramuar. So far no vehicles and research equipment have been procured nor any research officer been engaged, probably because the post has not yet been sanctioned. No research coordinator has been posted either.

It should be necessary to get a number of subordinate staff, trained for this purpose to form a useful organs of the research units.

It has been intended that the research coordinator, when recruited will establish linkages with various research institutions for research on biological, social and ecological matters. In this respect the Wildlife Institute of India will have to lay a key role. NGOs who are competent could be associated with socioeconomic research.

At present some local consultants and research organizations have been assigned with some works in B.T.R.

For intensive research two posts were sanctioned, one of an ecologist and another a Sociologist. At present a sociologist is in position. The ecologist worked for years has left B.T.R. No documents on any research work done by these researchers were available.

Monitoring and Impact Monitoring

Monitoring of changes is essential in a fast changing ecosystem of B.T.R. The changes in soil condition, vegetation quality and quantity, condition of water and air, condition of soil and vegetation due to grazing and fire, changes in faunal population, wildlife health and impact management practices need to be extensively and intensively monitored.

No report to study such changes, if any, was available for the purpose of co-relation with the present study. It is too early to expect any result from studies which have been undertaken lately.

Monitoring of impacts will enable the B.T.R. management to analyse and evaluate the change in B.T.R. brought about by various biotic and abiotic factors. The Management Plan desires monitoring at six months interval.

The B.T.R. has yet to evolve an elaborate systematic and management-oriented mechanism for impact monitoring.

The Management Plan suggests monitoring of :

- Physical changes of air, water (rivers, streams).
- The changes in vegetation.
- The changes in wildlife population.
- Wildlife health and diseases.
- Impact of management practices in and around B.T.R.
- The impact of Tourism.
- The impact of illegal grazing, pilferage, logging and harvesting of forest by fringe people.
- The ecodevelopment process.
- The activities of the members of FPC/EDC in relation illicit grazing, collection of NWFP, fuel wood and detection of poaching.

Evaluation of research projects every six months.

Supervision Mission observes that a consultancy for Impact Monitoring, which was pending with the Deptt. of Forests should be cleared. This should initiate a systematic monitoring and recording of impacts of FPC/EDC investments on Park protection. The PA staff requires to ensure that the Ecologist and Sociologist should closely be involved in monitoring work. Although it was management effectiveness using the recently published IUCN guidelines. This was being done.

The Mission stressed that establishment of a monitoring system should get the priority and that should be a part of the Management Plan.

The monitoring plan should be linked to and incorporate results from research studies and regular Park monitoring activities e.g. wildlife survey, etc. It should engage F.P.C./E.D.C. members in monitoring socioeconomic conditions, natural resources and fuelwood use, etc.

The monitoring efforts should help to determine remaining threats and whether the F.P.C./E.D.C. efforts are targeting the main users of and impactors on P.A. resources.

One officer should undertake over all responsibility for monitoring activities, maintenance of database and coordination with monitoring consultants.

This project component concentrates on using environment education and awareness and impact monitoring and research to develop effective support for the ecodevelopment project. It also concentrates on impact monitoring and research to improve understanding of issues relevant to the biodiversity conservation and then addressing these issues through P.A. management and encouraging positive interactions among P.A. authorities and local people.

Various activities include: developing environmental education and awareness strategy, improving visitor information and interpretation service, mass media campaigns, distribution of booklets, showing documentaries, developing and implementing visitor management and participatory tourism strategy.

Impact monitoring and research are seen as closely linked areas in this project. Impact monitoring could provide relationship between project activities and objectives and research is expected to throw light as to how this relationship could be made positive.

The ecological research could lead to identification of better ways of controlling fire, flood, wildlife depredation, understanding ethnobiology, indigenous resource management and preserving cultural heritage.

W.W.F prepared an awareness strategy. It is not known what was the reaction of the Forest Department. Some local NGOs are working together with B.T.R. authorities on this issue.

There is no special officer, for this purpose exclusively for B.T.R. Regular staff of B.T.R. do this job, The contribution of some clubs, especially of Alipurduar Nature Club is praise worthy.

The existing Nature Interpretation Centre, Ochid House, Wildlife Capture cages, nursery, etc. are commendable initial effort of the B.T.R.

Research should be conservation oriented and necessary for park management decisions. Research priorities have been identified in Management Plan, but the research work has not yet been shaped in B.T.R. but some have been prioritised.

Any research consultancy and/or study should contribute baseline information and monitoring and should be completed six months prior to the end of the project.

The researchers should involve local and tribal communities in research efforts, especially while monitoring use of NWFP, other PA resources.

It envisages fund allocation for research of about Rs. 1 lakh (for each) by the Field Director. This fund is to encourage staff and students of local institutions to undertake research relevant to Park Management and to build a local constituency for the Park.

Summary

These issues have been studied in details. It was concluded that the study should be situation oriented, adopted to local circumstances and responsive to local people and these should be action oriented and have a problem solving approach with stress on the environment education of the local people. However, these efforts are in the initial stages and no salutary effect was observed during the study.

Abbreviations

BTR	–	Buxa Tiger Reserve
NWFP	–	Non-Wood Forest Produce
NGO	–	Non-Govt. Organisation
FPC	–	Forest Protection Committee
EDC	–	Eco Development Committee
WWF	–	World Wildlife Fund
IUCF	–	International Union of Conservation of Nature
PA	–	Protected Area.

14

Protected Areas and
Sustainable Development
— Forester's Conservation Effort

This chapter records the findings of a study made by the authors in the Buxa Tiger Reserve, which could be a model for implementation of their observations in similar protected areas. They have also discussed about the various conceptions of world organizations in this regard.

The study methodology and findings were a role model in determining sustainable development relating to wildlife conservation, achievements, research, conservation strategy, policy on environment and development, legislation, institutional support and Government of India's silent revolution for environmental conservation.

Protected Areas and Sustainable Development
— Forester's Conservation Effort

In India the network of protected area consists of more than 80 National Parks, 441 Wildlife Sanctuaries covering more than 1,50,000 sq. km. of the Country. A National Coordination Committee has been setup to promote effective interdepartmental coordination for the control of illegal trade in Wildlife and Wildlife products in the country. Also an Interstate committee has been setup to review the Wildlife (Protection) Act, 1972, and their relevant laws.

The authors intend to discuss in nutshell various ramifications of Protected Areas (P.A.) and the measures taken by the Central and State Governments to protect them. The discussions will also cover the aspects of sustainability of development of the projects briefly.

At the outset it is necessary to define the terms Protected Areas (P.A.) and Sustainable Development (S.D.) which are intimately connected with each other.

Protected Area (P.A.) is defined by the Convention of Biological Diversity as "A geographically defined area which is designated or regulated and managed to achieve specific conservation objectives."

Other organizations, viz., The Fourth World Congress on National Parks and Protected Areas (1992), and Global Biodiversity Strategy (1992) also define the 'Protected Area' on same lines.

It's resources need wise maintenance and utilization by accurate inventory and taking protective measures. P.A.s contribute to the conservation of living resources and sustainable development by :

- Maintaining ecological processes and life supporting system.
- Preserving genetic diversity.
- Maintaining environmental stability to decrease intensity of flood, erosion, drought, grazing, fire and conserving water and soil and moisture.

- Maintaining productivity of ecosystem to conserve water and resources.
- Providing conservation education.
- Providing base for recreation and promoting of tourism.

Various strategies of P.A. remain incomplete unless the projects are sustainable; a sustainable development, however, has a complex definition as conceived by several authorities from the world. *The following is a summary showing the concepts of various authorities/ organizations from the world.*

Various Conceptions			Authorities
(a)	(i)	Eco-centric	Hugh Barton (2000)
	(ii)	Anthropocentric	... ...
(b)	(i)	People-centred (Social, Environmental and Economic)	W.C.E.D. (1987)
(c)	(i)	Maintenance of environmental quality.	Simon Bell *et al.* (1999)
	(ii)	Development that improves quality of life.	... ...
	(iii)	Economic development subject to maintaining the services and quality of natural resources.	... ...
	(iv)	Sustainability to be viewed in a holistic manner including economic, social and ecological components.	... ...
	(v)	Envisages imperative necessity of measuring the Biodiversity.	... ...
(d)		Sustainable Development indicates, social, economic, environmental facets.	United Nations (1991)
(e)	(i)	To maintain essential ecological processes and life support system.	W.W.F., I.U.C.N. and U.N.E.P. (1980)
	(ii)	To preserve genetic diversity.	... ...
	(iii)	To ensure sustainable utilization of species and ecosystem.	... ...
(f)		"Sustainable Development (is) a development strategy that manages all assets, natural resources, and human resources, as well as financial and physical assets for increasing long-term wealth and Well-beings.	Robert Repello (1986)

According to I.U.C.N. (1991) Sustainable Development implies increasing human productivity and the quality of the life while keeping within the carrying capacity of supporting ecosystem. Strategies for Sustainable National Development seek to improve and maintain well-beings of people and ecosystem.

Project Monitoring

Monitoring the projects were confined to look into the following aspects (only a few, mentioned among many)

- Improved P.A. management viz. Planning Organizational strength, Consolidation, Capacity building and Key ecosystem restoration issues.
- Village eco-development, viz. Mobilisation of awareness; Attitude of villagers, Staff, N.G.Os., etc.
- Nature education and awareness campaign for staff N.G.Os and general public.
- Impact monitoring and research to understand the issue and finding solutions.
- Analysis of the existing funding sources, financing mechanism in the project area and state.
- Analysis of the range of funding sources and financing mechanisms, including opportunities for recovering cost of selected services being used by different states and abroad.
- Consulting the Forest Department to make rough projection for post-project budget requirement to maintain and restore Biodiversity in the project area and continuing village eco-development initiative.
- Identification and meeting the key-stakeholders who could pro-actively promote the adoption of new funding sources and funding mechanism.
- Recommending the strategy on how the biodiversity conservation programme in the project can achieve financial stability.

Some Examples

Since the ratification of the Conservation on Biological Diversity in India in February 1994, several steps have been taken to meet the commitments under the convention as well as to bring the legislative, administrative and policy regime regarding biological diversity in tune with the Article of the convention to avail of the opportunities created by it. India has since participated in several International meetings concerning biodiversity. Twenty-ninth December is celebrated as the International Day for Biological Diversity to commemorate the date of entry into force of convention on Biological Diversity.

Under the Biodiversity Conservation, many projects are undertaken such as Wetland in States, Lake Chilka, Inventory of flora and fauna of some protected areas, Marine Parks, etc. The scheme on biodiversity conservation was initiated to ensure proper coordination among various agencies concerned with issues relating to conservation of biological diversity, and to review, monitor and evolve adequate policy instruments for the same.

Subsequent to the signing of the Convention of Biological Diversity by India during the UNCED Earth Summit (1992) extensive consultations were organized at various levels.

Biodiversity related activities were primarily concerned with very many conservation projects some of which are:

- Environmental Impact Assessment (EIA).
- Wetlands.
- Lake conservation.
- Biosphere Reserves.
- National River conservation.
- Ecodevelopment.
- National Parks.
- Tiger Reserves.
- Wetland fauna and several others.
- Impact of Ganga Action Scheme.
- Evaluation of Ganga Action Plan Works.
- Yamuna and Gomti Action Plan.
- Damodar Action Plan.
- National River Conservation Plan (NRCP).
- National Lake Conservation Plan (NLCP).
- National Afforestation and Ecodevelopment Board (NAEB).
- Integrated Afforestation and Ecodevelopment Project Scheme (IAEPS).
- Area Oriented Fuelwood and Fodder Project Scheme (OAFFP).
- Raising of Non-Timber Forest Produces (NTFP) including.
- Medicinal Plants.
- Monitoring and Evaluation of Afforestation Activities.

The Wildlife Institute of India has been enstrusted with the responsibility of acting as a nodal agency for research programmes under the Conservation of Biodiversity Component of Forest Research Education and Extension Project involving two sites namely the 'Great Himalayan National Park' in Himachal Pradesh and 'Kalakad – Mundanthurai Tiger Reserve' Tamil Nadu.

These are a few very broad outlines of some of many biodiversity projects running in India. In fact the foresters and administrators should have a clear idea about this multifaceted subject, 'Biodiversity', so that they may be fully aware of various imperative modalities of this subject while implementing them.

Wildlife Conservation Achievements have been a Special Issue

The following have been successfully initiated:

- A network of protected areas have been established.
- Wildlife Institute of India takes care of training and research in wildlife.

- Effective measures are taken to stop illegal trade in animals. (On the basis of the report of The Subramaniam Committee).
- A National Coordination Committee (NCC) has been setup to promote effective interdepartmental coordination for control of illegal trade in Wildlife and Wildlife Protection.
- An inter-State Committee has been setup to review the Wildlife (Protection) Act, 1972 and other related laws.
- The Indian Board for Wildlife (IBWL) the apex advisory body in the field of Wildlife Conservation has been reconstituted.
- Assistance was provided to the States for taking-up programme of Ecodevelopment around National Parks and Sanctuaries indicating the higher resources in order to achieve ecologically sustainable economic development of these areas.

Project Elephant

This project was launched in 1991-92.

- For ecological restoration of natural habitat.
- For Eco development.
- Conservation of habitat.
- To mitigate man-elephant conflict.
- To preserve migratory routes.

Research Projects

Some of the projects undertaken are :

- All India Coordinated Research Project on Aero-allergens and Human health.
- Coordinated Research Project on Ethnobiology.
- Inter-Government Panel on Climate Change (IPCC).
- G.B. Pant Institute of Himalayan Environment and Development.
- Research on Biosphere Research, Wetlands and Mangroves.
- Indian Council of Forestry Research and Education. (ICFRE), Dehradun.

Project Tiger

This project was launched in 1973 and remarkable success was achieved –

- to ensure maintenance of a viable population of tiger in India for scientific, economic, aesthetic, cultural and ecological values.
- to preserve for all times, areas of such biological importance as a national heritage for the benefit, education and enjoyment of the people.

There are 23 projects covering 33046 sq.km. all over the country.

Protected Areas Vis-a-vis the Role of Government

The ministry of Environment and Forests, Govt. of India is the nodal agency for the planning, promotion and coordination of environmental and forestry programmes.

The primary tasks involve –

- Conservation and Survey of Flora and Fauna.
- Conservation and Survey of Forest and Wildlife.
- Prevention and Control of Pollution.
- Afforestation and Regeneration of degraded forests.
- Assessment of Environmental Impact.
- Implementing Environmental and Forestry Programme.
- Promotion of Environmental and Forestry Programme.

National Conservation Strategy and Policy Statement on Environment and Development

The National conservation strategy and Policy statement on Environment and Development 1992, the National Forest Policy, 1998 and Policy statement for abatement of pollution, 1992 are the major policy instruments of the Government for dealing with various facets of environment and development in a comprehensive manner.

Biosphere Reserve

The Biosphere Reserve Programme is a pioneering effort at pursuing the increasingly difficult yet urgent task of conserving Ecological diversity under mounting pressure. It is to protect Representative ecosystem. 12 Reserves have been formed so far.

Wetland (National Lake Conservation Plan)

The following are some of wetlands that have received attention:

Kolleru, Wulan, Chilka, Loktak, Bhoj, Sambhar, Pichola, Sasthamkolla and Ashtamudi, Nalsarobar, Hariki, Uyni, Renuka, Pondam, Chakatratal, Kabar, East Calcutta, Sukna, Deeper Beel among several others.

Legislation and Institutional Support

Legislation

Since the enactment of the Environmental (Protection) Act, 1986, several actions have been taken by the Ministry of Environment and Forests to create a comprehensive legal and institutional infrastructure for safeguarding the environment. These include framing of rules, notification of standards, notification of environmental laboratories, delegation of power, identification of agencies for management of hazardous chemicals, setting-up of Environmental Protection Council in the States, etc.

The following were enacted:

- The National Environment Tribunal Act, 1995.
- Amendment of Water (Protection and Control of Pollution) Cess Act, 1977.

Government of India's Silent Revolution for Environmental Conservation

In connection with P.A.'s the G.O.I. has summarised in their achievement in the report. "A Silent Revolution for Environmental Conservation" 1999. Some salient features of this report is presented here in after:

Environmental Action Plan for Controlling Pollution at the National Level

For this the Ministry prepared a draft on National Environmental Action Plan.

 (i) Industrial Pollution
 (ii) G.O.I. has listed 1551 industrial units in the highly polluted category
- Use of low ash coal by Thermal Power Plants.
- Environmental Action Plan for control of pollution at religious places.
- Taj Protection Mission.
- Crackdown on Polluting Industries.
- Pollution control in Delhi.
- Sustainable Development strategies.
- Hazardous substance management.

Vehicular pollution
- Condemnation of old vehicles.
- Use of LPG in place of petrol/diesel.
- Induction of antipollution mechanism in the vehicles.

(iii) Environmental Impact Management
- Monitoring of Development projects.
- Constitution of Authorities.
  (Coastal Zone Authority, Aquaculture Authority, Central Ground Water Authority).
- Restoration of mining areas.

Urban Aesthetics
— Ecofriendly Roles of Parks and Gardens

It is a vast subject on which the authors could prepare a voluminous treatise, but the subject being a small part of this treatise they have discussed the salient points very briefly.

15

Urban Aesthetics
— Ecofriendly Roles of Parks and Gardens

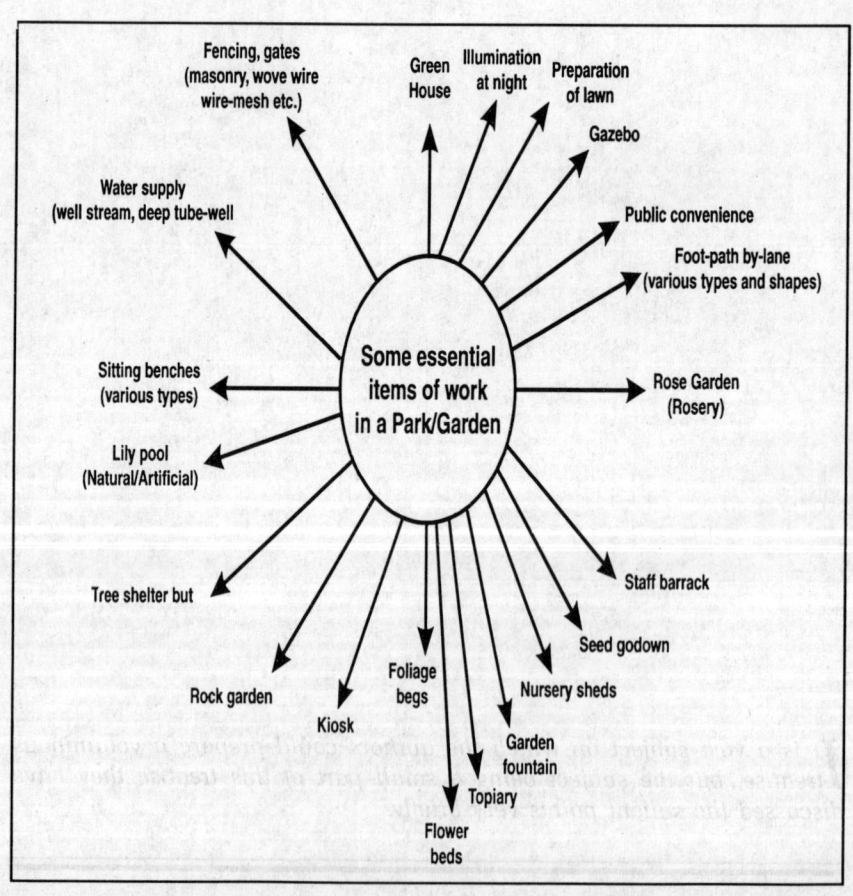

These are only a few broad items of work in a park to be undertaken by a park manager.

The coauthor D.D. Sarkar and Late Amal Lodh, Forest Ranger who shaped the urban parks and gardens of West Bengal activities for some years drew realistic estimates on the following items of work so that nurserymen, and the managers of private gardens may draw cash estimates according to local existing rates.

These will testify to the fact that foresters were alive to the need of urban people, especially of the park and garden managers.

Parks and Gardens, Wing of the Department of Forests was shaped by an eminent forester in West Bengal Late A. B. Guha Neogy, I.F.S. (Retd.) the first special officer of the organisation in early 1970s. His golden fingers created a series of flower and foliage gardens, avenues flanked by flowering shrubs and trees and nurseries. The wave of success in various parks and gardens created an aesthetic ethics in every creek and corner of the country. This aesthetic revolution may be noticed in mushroom growth of hundreds of nurseries in the vicinities of Calcutta and also around many other cities of India especially the Southern India.

The aesthetic wave has crept into every house, school, college, office, hotel, restaurant and various rooms, corridors and premises which are being decorated by colourful flowers and foliage.

The efforts of Late Guha Neogi were very successfully intensified by his successors Late J. R. Sen, I.F.S. (Retd.) and Shri U. Dasgupta, I.F.S. (Retd.).

There has been increasing demands for forest plants and fruits to be used as decorative material. We are in an era where no plants are to be considered as weeds as each has a role to play as decorative material. Several species are being over exploited, signifying a warning of over exploitation.

16

Moratorium
— Why A Moratorium?

The authors have suggested 15 years 'moratorium' to build a strong foundation on which the country's superstructure of modern India should be built. They feel that India should cut-off all extravagant expenditures of colourful ceremonies, film shows, beauty contests, car-racing, expensive rituals, luxurious travels, savoury cuisines, etc. for the time being. They have mentioned the achievements of Wangar Mathai of Kenya, F. Roosevelt, ex-President of USA, the valuable remarks made by Dr. A.P.J. Abdul Kalam, President of India, on the eve of Independence Day, 2005, and the observations of B.B. Vohra, ex-Chairman, Flood Commission of India. They have recommended creation of plantations over vast wastelands and all areas otherwise unsuitable for other purposes. The authors are convinced, as the aforesaid authorities are, that mass-scale afforestation only can save India from a major catastrophe.

16

Moratorium
— Why A Moratorium?

If India has to survive and overcome the present environmental and landuse dilemma, and save itself from the inevitable doomed destiny it has to observe a moratorium to halt extravagant expenditures in various public and private ceremonies. This moratorium has to last at least for 15 years.

The pattern of changes suggested involves life style, ceremonial expenses, school and college curricula, practical training to all service personnel, gay entertainments, whole night rendezvous in hotels and restaurants and similar other expenditures. The huge sum of infructuous money misspent in such food, fashion, fancy movies, so called creative and colourful madness, extravaganza in various campus with and stunning moves and rocking rhythms may be saved and spent for the purposes now suggested. Let uncreative music, fashion, poetry and dances extravaganza be dumped in a basket for sometime.

All planning efforts to build up a better India will not yield any fruitful result due to the poor and unstable foundation on which our country stands. Long ago Swami Vivekananda visualised that which our youth should possess strong will and determination, based on 'man-making and character-building' system of education; besides the villagers must be capable of improving their own fate by hard work, not depending on outside help. What we find today, even after more than half a century of independence, that the dream of the sage of the sages – **Swami Vivekananda** remains unfulfilled.

India needs adoption of 15 years' 'moratorium' to build a strong foundation on which the country's superstructure of modern India can be built; for this India needs :

- Cutting off all extravagant expenditures of colourful ceremonies, film shows, beauty contests, car racing, expensive rituals, luxurious travels, building luxurious hotels, savoury cuisine, and spending heavily on entertainments.

- School/College curricula must contain syllabus highlighting the catastrophic landuse situation of the country and need for massive watershed management, afforestation and soil conservation works. The students should know about the socioeconomic backwardness, political instability, poverty, illiteracy, primitive social customs, drought, flood, etc. of the country and the usefulness of forest environment to mitigate those hazards. The entire education system should be based on the 'man-making and character building' – system of education besides some fields-based works.

- An effort similar to Wangar Maathai of Kenya to protect forests (this fetched for her Peace Nobel Prize of 2004) is necessary to save the country.

- India has to build up a department on the line (C.C.C) Civilian Conservation Corps which the late president F. Roosevelt of USA conceived in 1931 to tackle the problems of unemployment, erosion, drought, wind storm, soil wash by massive afforestation and soil conservation programme.

- This suggestion gets support in President of India – Dr. A.P.J. Abdul Kalam's Independence Day (2005) deliberation for an action plan for rural employment generation to help achieve 10 percent G.D.P growth rate. Dr. Kalam said (based on Planning Commission figure) that the objectives will be achieved through employment generation of 76 million people in rural areas for which he suggests Bio-fuel plantation, waste land development, water harvesting, bamboo plantation, and the like.

The people should temporarily forget the comfort of luxurious extravaganza of five and seven star hotels, multiplex, sophisticated and expensive private and official functions, abandon cheap pop-song, burning of petrol/diesel in activities other than essential services, more liquor shops, biting Mughlai, Thai, Afghan, Chinese, Tibetian and Continental cuisine and sucking guzzler's delight, Bratwurst (Chicken Pork Sausages), Barthend (Chicken Roast), leberkaese (meat loaf with fried onions and eggs) and the like.

Even at this crucial sociopolitical and socioeconomic unstable situation of the country a good section of intelligentsia (connoisseur of culture of the country) dream of whole night rendezvous round the year, the sights and sounds of city's festive fever, the flavours of thousand savoury and sweet delight, roam about from pandal to pandal in quest of cultural evolution, prefer fun and frolic, being oblivious of critical need of the country for which Mother India is in need of a great sacrifices. The entire length of country's border is teeming with criminals, local gangsters and intrusion and the country faces multifaceted crisis.

At this juncture afforestation, watershed management and soil conservation issues should be considered as National issues and must be tackled on a war footing. Our school and college students are environment conscious on paper only and they should actively participate in such programmes.

Some experts feels that the trend of environmental degradation indicates the several departments should be brought under one umbrella and a firm policy and action plan developed, followed by a strong political will to save the country from present disaster.

Planning Commission of India (1982) indicated that India can achieve ecological security by increasing vegetal cover to tackle the problem of serious degradation in various fields.

The authors sincerely feel that Indian Foresters are sure to take a leading role in shaping the country as proposed as they have 200 years' tradition in conservation and afforestation work.

B. B. Vohra, ex-chairman of Flood Commission of India, in his report suggested that there should be a Ministry of Land Management and creation of a National Land Development Board to finance various land improvement projects. Mobilization of students should be most useful for large scale soil conservation and afforestation programmes. There are immense scope of voluntary organizations to involve themselves in these activities. Vohra identifies renovation of India's 5,00,000 odd tanks and 900 odd project reservoirs and denuded watershed of India's flooded rivers.

Mention may be made of ITC's one of the Social Forestry Programmes in Chetavarigudam village of Khammam district of Andhra Pradesh, where it used high-yielding disease resistant, clonal saplings developed by it at its state-of-the-art biotechnology research centre. More than 26,500 hectares of wastelands had been rejuvenated with 108 million saplings, creating livelihood opportunities for nearly 2,00,000 people. This gives a positive guide to the future livelihood and employment of a vast population to be engaged in afforestation and soil conservation works all over the country.

The citizen's promise of the millenium should be:

- To create a silvan landscape and golden future;
- to fulfil the oath to plant up millions and millions seedlings of foliage, flowering and timber plants to conserve and enhance the biodiversity; and
- to protect millions of people from environmental hazards by planting billions and billions of seedlings.

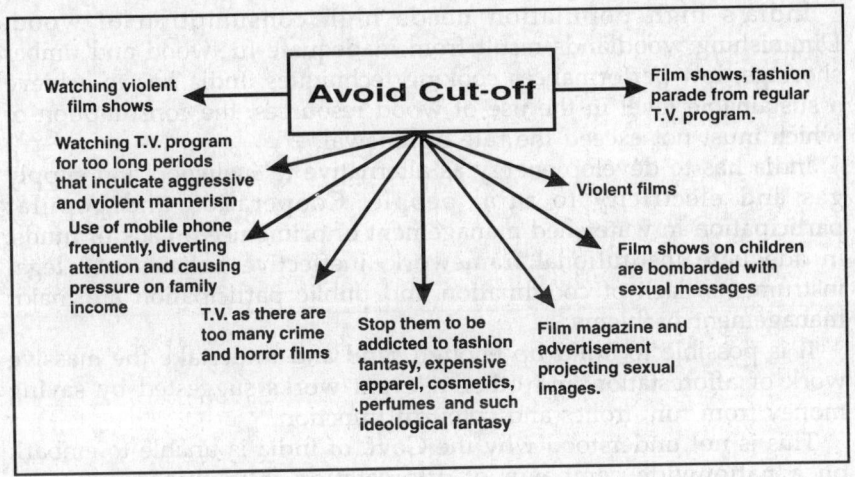

Avoid Cut-off

- Watching violent film shows
- Watching T.V. program for too long periods that inculcate aggressive and violent mannerism
- Use of mobile phone frequently, diverting attention and causing pressure on family income
- T.V. as there are too many crime and horror films
- Stop them to be addicted to fashion fantasy, expensive apparel, cosmetics, perfumes and such ideological fantasy
- Violent films
- Film shows, fashion parade from regular T.V. program.
- Film shows on children are bombarded with sexual messages
- Film magazine and advertisement projecting sexual images.

MORATORIUM VIS-A-VIS CHILDREN'S
do's and don'ts

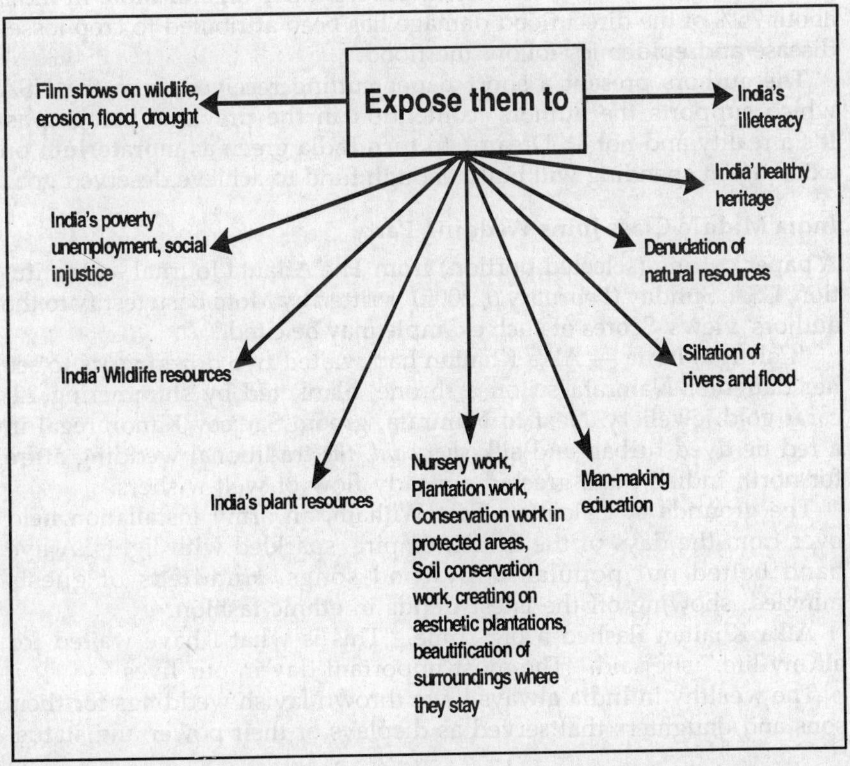

Expose them to

- Film shows on wildlife, erosion, flood, drought
- India's poverty unemployment, social injustice
- India' Wildlife resources
- India's plant resources
- Nursery work, Plantation work, Conservation work in protected areas, Soil conservation work, creating on aesthetic plantations, beautification of surroundings where they stay
- Man-making education
- India's illeteracy
- India' healthy heritage
- Denudation of natural resources
- Siltation of rivers and flood

India's high population needs high consumption of wood. Diminishing woodlands result from inadequate fuelwood and timber shortage, low-performances cooking techniques. India has to achieve a sustainable level in the use of wood resources, the consumption of which must not exceed the rate of renewal.

India has to develop energy as alternative to fuelwood and supply gas and electricity to rural people. Cooperation and popular participation in watershed management of prime need. Lack of funds, in adequate institutional framework, ineffective policies and legal instruments, lack of coordination and public participation are major management problems.

It is possible to build up enough fund and undertake the massive work of afforestation and other relevant works suggested by saving money from fun, frolics and colourful function.

This is not understood why the Govt. of India is unable to embark on a nationwide campaign of afforestation involving community participation as by the Govt. of Kenya.

United Nations Environmental Programme (UNEP) is concerned with land degradation national soil policy for protection of soil resources, conservation of biological diversity and depletion of Forest water resources. India may seek advice from this organization: In India about 75% of the direct flood damage has been attributed to crop losses disease and epidemics follow the flood.

The authors present a short paper cutting received from the USA which supports the authors' contention in the previous paragraphs. It's a reality and not a "Dream" to turn India green as **moratorium** on extravagant spending will build enough fund to achieve deserved goal.

India Middle Class Joins Wedding Party

A paper cutting (selected portion) from The Atlanta Journal – constitution, USA, Sunday (February 6, 2005) written by Moni Basu testify to the authors' views. Scores of such example may be cited.

"Calcutta, India — Alka Khaitan had waited two dozen years to see her daughter Namrata sit on a throne, blanketed by shimmering 22-carat gold jewellery. Next to Namrata, groom Sanjeev Kanoi, regal in a red tie-dyed turban and silk *sherwani, the* traditional wedding attire for north Indian men, greeted a steady flow of well-wishers.

The grounds of Calcutta's Fort William, an army installation held over from the days of the British empire, sparkled with lights. A live band belted out popular Bollywood songs. Hundreds of guests mingled, showing off the latest trends in ethnic fashion.

Alka Khaitan flashed a big smile. "This is what I have waited for all my life," she said. "The most important day in our lives."

The wealthy in India always have thrown lavish weddings for their sons and daughters that served as displays of their power and status.

But India's booming economy has handed a great deal more buying power to the country's burgeoning middle class and extravagance is no longer reserved for the elite.

Urban weddings start at $15,000 and rise to as much as $10 million, which is what a prominent Indian businessman shelled out for recent nuptials for one of his children. It's another example, said economist Ashok Mitra, of the widening gap between India's rich and poor in a land where the average yearly per capita income is still only $600.

The Indian wedding industry, valued at a whopping $10 billion, churns into action in October but the highest number of marriages take place in the astrologically auspicious months of November, January and February.

Good days to circle the matrimonial fire, as is Hindu custom, are determined by the celestial charts. On one especially favourable day last November, New Delhi alone saw as many as 14,000 weddings.

Cashing in on Love

In Calcutta wedding venues are fully booked, and streets are clogged with wedding processions — grooms often ride to the venues on a decorated horse. Rooms in the nation's luxury hotels are hard to come by. Billboards and advertisements for silk wedding saris and gold and diamond jewellery are as common as ads for cellphones and movies. Calcutta has even sprouted the nation's very first mall dedicated entirely to the business of marriage.

"It's like one-stop shopping for weddings," said sari vendor Suresh Kant Ojha.

The most expensive sari in the store can set a customer back by nearly $10,000, he said. He pulled out a ghagra (long, flared skirt) and took six months to sew and weighed more than 10 pounds because of the intricate gold brocade work.

"People are celebrating weddings like a big festival these days," said Rajiv Kanodia, director of Shagun, the wedding mall in the hart of the city. "They have the money and they are spending it. So we had this idea to open a wedding mall, and so far business has been booming. In fact, people are using up their entire life savings on weddings."

People like Irfan Ali, 24. The software engineer lives and works in Sydney, Australia, but came home to Calcutta to get married on Valentine's Day to an Indian woman from Canada.

At the wedding mall, Ali tried on a series of ethnic shirts, pants and jackets. He said he would purchase one outfit for each of the eight events planner before and after the actual ceremony.

Ali said he has invited more than 5,000 people to the wedding. His parents will pay more than $135,000 for the event.

Kanodia said parents are eager to ensure a dazzling trousseau for their daughters. A bride can wear as many as 60 items just during the wedding ceremony.

More than anything, said Kanodia, weddings have become a sport of one-upmanship.

No Longer Simple

It starts with the invitation, often a multitiered assortment of bejewelled cards that used to be accompanied by traditional fruits and nuts. These days, the ultra-rich send invitations with gifts of Prada bags or Hermes scarfs. Sometimes, India's most famous artists, are commissioned.

Guest lists can resemble a who's who of politicians and celebrities, none of whom may actually know the bride, the groom or their families.

In Calcutta, the upper echelons of the ruling Communist Party are often seen at the city's glitziest weddings. Newspapers and magazines devote an enormous amount of ink to the wedding of the day.

Gone, too, are the days of a simple sit-down dinner. Indian weddings feature as many as eight or 10 food stalls that feature everything from tandoori to tiramisu.

At Namrata Khaitan's wed ding, guests used four or five plates each to sample the North Indian, South Indian, Italian and Chinese menu items and a dozen desserts.

"I have only one daughter," said Alka Khaitan. 'I have saved money for her wedding from the day she was born. That"X throw her a big wedding is even more important to me than her education."

This single instance is enough to uphold the authors' suggestion for a moratorium and create a massive fund.

India needs a massive fund for it faces further serious crisis as detailed below and only massive green cover and integrated soil work may be able to combat the present and incoming disaster.

"Tsunami" and "El-Nino"

Recent disaster due to "Tsunami" of December 26, 2004 resulting in a massive loss of life and property is an indication of India's poor, disaster control measure. Indication of drought in Orissa having an "El-Nino" influence is a severe and perhaps a realistic warning that India needs more and more vegetal cover and soil conservation engineering works.

Himalayan Squeeze on Rivers of Life

It's a paper cutting from the Telegraph (dt. 15/03/2005) on global warming and its effect on India. The facts mentioned in the paper uphold the author's conception of a 'Moratorium' as it is the massive afforestation and soil conservation measures that can combat the anticipated disasters; and here comes the active participation of million of Indians for which adequate fund necessary for this purpose will accrue from moratorium as intended in this treatise.

After the sea, it is probably the turn of the mountains to wreak havoc.

The Himalayan glaciers are receding at among the fastest rates in the world due to global warming, threatening water shortages for millions of people in India, China, Nepal, a leading conservation group has warned.

The Himalayan glaciers feed seven of Asia's greatest rivers (see chart). The glaciers ensure a year-round supply of water to hundreds of millions of people in China and the Indian subcontinent.

The Worldwide Fund for Nature (WWF) said the Himalayan glaciers are receding 10 to 15 metres per year on average that the rate is accelerating as global warming increases.

In India, the Gangotri glacier is receding at an average rate of 23 metres per year, the study said.

"The Himalayan glaciers are amongst the fastest retreating glaciers globally due to the effects of global warming," the WWF said in Geneva.

"This will eventually result in water shortages for hundreds of millions of people who rely on glacier-dependent rivers in China, India and Nepal," it said.

"The rapid melting of Himalayan glaciers will first increase the volume of water in rivers, causing widespread flooding," said Jennifer Morgan, director of WWFs global climate change programme.

"But in a few decades this situation will change and the water levels in rivers will decline meaning massive economic and environmental problems for people in western China, Nepal and northern India," she said.

The WWF released the study before a two-day ministerial roundtable in London this week of the 20 greatest energy-consuming countries, to be followed by a G8 meeting focussing on climate change in Africa. Such meetings have assumed an added significance against the backdrop of the tsunami that ravaged several countries in December.

The WWF called for work towards reducing carbon dioxide emissions – blamed for global warming – plus increasing the use of renewable energy and energy-saving measures.

The countries participating in the Energy and Environment Ministerial Roundtable include Canada, France, Germany, Italy, Japan, Russia the UK and the US plus non-G8 countries Australia and Brazil.

"It is a fact that the Himalayan glaciers are receding at a very fast rate. Even the working group of Inter Governmental Panel on Climate Change has recognised the phenomenon sometime back," Suruchi Bhadwal, an Indian researcher working on the impact of climate change, told *The Telegraph*.

Sugato Hazra, a scientist from the oceanography department of Jadavpur University, said the findings of a study with which he is associated support the WWF warning.

"We have been mainly working in the Sunderbans area to trace the environmental impact of climate change and our findings show that both the temperature at the sea surface as well as the sea levels are consistently rising," Hazra said.

"The temperature at sea surface is increasing 0.019 degree every year – about one degree over 50 years – while the sea level is gaining a height of 3.14 millimetre a year," the scientist added.

India has to take all possible measures to combat these issues with topmost priority to save the country from catastrophes.

17

An Overview
— Some Salient Points for Considering Building a Better India

*T*he need for saving India's dwindling forest cover has prompted Indian Foresters to under go management courses. Forest management is all about conservation and maintenance of forest wealth and to administer the forests. The work implies application of technical forestry principles, practices and business techniques, hence the forest management involves the practical application of ·scientific, economic and social principles of forests. Foresters manage a forest, keeping some multifaceted objectives in view some of which are to maximise production of timber, fuelwood, furniture wood, paper and pulpwood, fodder, and to meet the local demand of the people, exploitation, creation of plantation and also management of wildlife. Every state has its own management organization and the forest policy provides conceptual guidelines on forestry operations. The administration includes analytical and managerial skills, application of concepts to managerial problems, synthesis, technical skills, fieldwork, basic accounting, management accounting, financial management and marketing and project management. Forest management in fact means the accomplishment of multiple goals such as sustainable forest management, livelihood issues in rural India, community forestry, gender issues, legal policy analysis in forests, protected area and biodiversity conservation, management of ecosystem and watersheds, soil and water conservation, maintaining biodiversity and hydrological cycles and carbon sequestering. So the management issue involves judicious combination of management, service and forestry sciences besides several others.

In the event of perilous landuse situation worsened by forest depletion, fire damage, grazing menace, encroachment, illicit felling which started a thousand years back and still continuing, have made the task challenging.

The authors have presented these issues in various chapters and have drawn an outline to turn the country green in a span of 15 years.

17

An Overview
— Some Salient Points for Considering Building a Better India

India's forests meet a part of the demand of it's vast population of fuel, fodder, wood besides providing environmental services and watershed protection. But the degradation of forests continues and the present 0.08 ha. of per capita forests represent poor quality and resource.

Therefore, for sustainable management of India's forests a new strategy is needed. The Government has lately decided to make villagers and vicinity people as partner in forest management, adopting modern technologies to ensure effective implementation of forest policies, setting development priorities based on the cost and benefits.

The social forestry and watershed development activities met with mixed success. So a change was necessary for lessening the burden of heavy population (more than a billion) creating heavy pressure on the existing forests. The demand was high and the forest were being over-exploited. A vast livestock population has caused severe damage to whatever little regeneration is there.

The private sector has not contributed to developing the forest sector. Farm forestry development has considerable potential. Afforestation programme on public and private lands, indicates adoption of improved technology. The most important gaps being the quality of planting materials; lack of appropriate models and modalities for generation of degraded forests with people's participation: planting practices and range of models for commercial and form forestry production; and a weak information transfer process. Women's participation has not yielded much success.

The most important forest policy goal for future would be to improve—

• Forest protection and
• Forest management methodologies.

Here traditional policy may not be effective to prevent degradation.

At the this level the Government has enlisted the cooperation of local population and other agencies.

There are several inherent conflicts between users of forests which are not enumerated here. The concept of rotational grazing and increasing fodder production still remains on paper initiated 70 years back. Institutional linkage between forest departments need tribal development authorities has not yielded useful solutions.

So the strategy now should be to :–

- improve incentives for local population and private development
- improve the technological means to solve production from forests (better seeds, nursery and management)
- increase the effectiveness of public sector forest managements and development (N.G.Os)
- prioritising areas to protect and develop and
- topics that requires further analysis improve (data base, etc.)

Some Facts

- Social forestry undertaken between 1975-90 led the planners and donor agencies to shift attention from non-forests to forest land due to mixed results.
- In 1988 the Government of India revised the old timber-oriented forest policy and requested the state governments to create subsistence requirements of the forest dwellers as the initial charge on forest produce.
- In June, 1990, the Government of India recommended to various state governments to encourage the involvement of village communities and voluntary agencies in the protection and management of forests through joint Forest Management (JFM) was never considered as panacea for deforestation or for alleviation of rural poverty. JFM was just to halt land degradation.

So the strategy for rehabilitation was considered to be multi-pronged, some of the key issues are as follows.

Forest Land

- Comparative advantages in planting on revenue land and forest land of which the latter land is preferred.
- Initiating of Farm forestry and Agroforestry programmes.
- Change in silviculture practices with the changed requirement. Stress to be laid on Non-Timber Forest Produce (NTFP).
- Proper choice of species.
- Notice board to notify the type of work the coupe will be allotted to.
- Changes suggested in states Forest Act in view of Government of India's, Forest Conservation Act.

Revenue Land

- Lending of technical support to all planting on revenue land.
- Grazing land should be identified.
- Model afforestation schemes to be prepared for implementation by the Panchayats.
- Development of fodder grass to be encouraged.
- Degraded land to be afforested.

Farm Forestry

- Planting to be linked with industries.
- No subsidy to be given.
- Short rotation species are to be selected.
- Proper nursery are to be made.

People's Participation and Joined Forest Management (JFM)

The Forest Service in India has undergone great changes in last ten years. The foresters have been exposed to new ideas who are working with the people.

More radical orientation towards giving property rights to the communities and accepting a reduced role of a facilitator for the Forest Service are being thought.

- NGOs and voluntary organizations should be associated to provide an interface with local people to promote joint forest management programme or to settle forestry related disputes.
- Involvement of women and women organization.
- People's participation should be prompted by enabling the people to have usufruct of intermediate and a definite shares on final produce.

Bibliography

1. Ali Salim (1941)   *The Book of Indian Birds*.
2. Anon (1953)   *Final Report of All India Soil Survey Scheme*, ICAR.
3. Baker, E.C.S. (1930-'35)   *The Fauna of British India*.
4. *Biodiversity Assessment* (1996)   H.M.S.O., London.
5. Brandis, D. (1884) The Progress of Forestry in India, *Indian Forester* 10 (10).
6. Champion, H.G. and Seth, S.K. (1968)   *Revised Forest Types of India*.
7. Chaudhuri, A.B. (1985)   *Environment and Resources of Tropical & Temperate Forests*.
    Chaudhuri, A.B. (1993)   *Trees and the environment*.
    Chaudhuri, A.B. (1996)   *Wetland Ecology*.
    Chaudhuri, A.B. (2004)   *Plants of Darjeeling and Jalpaiguri*.
8. Chaudhuri, A.B. and Sarkar, D.D. (2002)   *Biodiversity Endangered*.
    Chaudhuri, A.B. and Sarkar, D.D. (2003)   *Megadiversity Conservation*.
    Chaudhuri, A.B. and Sarkar, D.D. (2003)   *Wildlife and Ground Flora*.
    Chaudhuri, A.B. and Sarkar, D.D. (2004)   *Project Tiger Reserves*.
    Chaudhuri, A.B. and Sarkar, D.D. (2002)   *Golden Triangle of Ecoeducation Tourism*.
9. Curtis, S.J. (1993)   *Working Plans for Sundarbans Forest Division*.
10. Forest Centenary (1964)   *100 Years of Forestry in India* (1864-1964).
11. Parker, R.N. (1923)   *The Indian Forest Department*.
12. Puri, G.S. (1960)   *Indian Forest Ecology*.
13. Ribbentrop, B. (1900)   *Forestry in British India*.
14. Stebbing, E.P. (1922)   *The Forest of India* - Vol-I, London.
    Stebbing, E.P. (1924)   *The Forest of India* - Vol-II, London.
    Stebbing, E.P. (1926)   *The Forest of India* - Vol-III, London.
15. Shebbeare, Z.O. (1930)   Fire and Sal, *Indian Forester* 56(3).
16. Sivaramakrishnan, K. (1999)   *Modern Forests*.
17. Sagreya, K.P. (1967)   *Forests and Forestry*.
18. Schleich, W. (1884)   Review of Forest Administration for British Indian, *Indian Forester* 10(8).
19. Troup, R.S. (1905)   *Fire Protection in Teak Forests of Burma*.
    Troup, R.S. (1921)   *Silviculture of Indian Trees*.
20. *Wildlife (Protection) Act*, 1972   Govt. of India

Index